T0259704

Dynamics and Control of Lorentz-Augmented Spacecraft Relative Motion

Ye Yan · Xu Huang · Yueneng Yang

Dynamics and Control of Lorentz-Augmented Spacecraft Relative Motion

Springer

Ye Yan
College of Aerospace Science and Engineering
National University of Defense Technology
Changsha, Hunan
China

Yueneng Yang
College of Aerospace Science and Engineering
National University of Defense Technology
Changsha, Hunan
China

Xu Huang
College of Aerospace Science and Engineering
National University of Defense Technology
Changsha, Hunan
China

ISBN 978-981-10-9665-5 ISBN 978-981-10-2603-4 (eBook)
DOI 10.1007/978-981-10-2603-4

© Springer Science+Business Media Singapore 2017
Softcover reprint of the hardcover 1st edition 2016
This work is subject to copyright. All rights are reserved by the Publisher, whether the whole or part of the material is concerned, specifically the rights of translation, reprinting, reuse of illustrations, recitation, broadcasting, reproduction on microfilms or in any other physical way, and transmission or information storage and retrieval, electronic adaptation, computer software, or by similar or dissimilar methodology now known or hereafter developed.
The use of general descriptive names, registered names, trademarks, service marks, etc. in this publication does not imply, even in the absence of a specific statement, that such names are exempt from the relevant protective laws and regulations and therefore free for general use.
The publisher, the authors and the editors are safe to assume that the advice and information in this book are believed to be true and accurate at the date of publication. Neither the publisher nor the authors or the editors give a warranty, express or implied, with respect to the material contained herein or for any errors or omissions that may have been made.

Printed on acid-free paper

This Springer imprint is published by Springer Nature
The registered company is Springer Nature Singapore Pte Ltd.
The registered company address is: 152 Beach Road, #22-06/08 Gateway East, Singapore 189721, Singapore

To my wife, Q. Huang, and my daughter, J. Yan

Ye Yan

To my parents, G. Huang and X. Ling, for their love

Xu Huang

To my parents, Z. Yang and A. Lin, my wife, J. Sun, and my daughter, Z. Yang

Yueneng Yang

Preface

Lorentz spacecraft is a new conceptional space vehicle. It actively generates electrostatic charge on its surface to induce Lorentz force via interaction with the ambient planetary magnetic field. The induced Lorentz force could be used as propellantless magnetic propulsion for orbital maneuvering. Traditional spacecraft rely on the thrusters to perform such maneuvers, and the maneuvering capability is strongly restricted by the amount of the available propellant onboard. Differently, the conception of Lorentz spacecraft provides a propellantless means to solve this problem, thus greatly enhancing the maneuvering capability and extending the mission duration. Due to this compelling advantage in saving fuels, dynamics and control of this new vehicle have raised great research interests.

This book aims to investigate the dynamics and control of Lorentz-augmented spacecraft relative motion. It consists of three parts, including introduction, theoretical analysis, and applications. The first part (Chap. 1) gives a detailed review on the theoretical and experimental developments of Lorentz spacecraft. Also, this chapter introduces three types of typical space missions. That is, spacecraft hovering, formation flying, and rendezvous, which all belong to the category of relative orbital motion. The second part (Chap. 2) develops theoretical models of Lorentz-augmented spacecraft relative motion about arbitrary elliptic orbits in both two-body and J_2-perturbed environment. Also, these models explicitly include the magnetic dipole tilt angle, which are more representative of the Earth's magnetic field, and are expected to derive more precise results. The third part (Chap. 3, 4, 5 and 6) proposes detailed relative navigation and control strategies for Lorentz-augmented space missions. Two navigation schemes for Lorentz spacecraft relative motion are introduced and compared in Chap. 3. The remaining three chapters elaborate the control schemes for Lorentz-augmented spacecraft hovering, formation flying, and rendezvous, respectively. Both optimal open-loop and closed-loop controllers are designed for these typical applications.

We have conducted the related research works since 2012. I wish to acknowledge my Ph.D. Student, Xu Huang, for his concentrated devotion in this study. His creative thinking and persistent diligence impress me very much. I particularly appreciate his contributions to the fulfillment of this book. Also, special thanks to all the scientists and researchers who preceded us in the theoretical and experimental developments of Lorentz spacecraft.

Furthermore, this work was supported in part by the Fund of Innovation of the Graduate School of the National University of Defense Technology under Grant B140106, and in part by the by the Hunan Provincial Innovation Foundation for Postgraduate under Grant CX2014B006.

Changsha, Hunan, China Ye Yan

Contents

Chapter 1
Introduction

1.1 Background

Traditional orbital maneuvering is mainly propelled by thrusters using chemical fuels on board [1]. Therefore, the duration of space mission is generally constrained by the amount of the propellant on board [2]. Once the propellant depleted, the maneuvering capability of the spacecraft will basically be lost. To get rid of this constraint, an advantageous and preferable alternative is to use propellantless propulsion. Novel means of propellantless propulsion include the geomagnetic Lorentz force [3], inter-spacecraft Coulomb force [4], and solar radiation pressure [5], and so on.

An electrostatically charged spacecraft is subject to the geomagnetic Lorentz force as moving in the planetary magnetic field. The induced Lorentz force could be used as propellantless propulsion for orbital maneuvering. Such novel space vehicle is referred to as the Lorentz spacecraft. Notably, due to the inherent physical mechanism, the induced Lorentz force could only act in a direction perpendicular to the local magnetic field and the vehicle's velocity relative to the local magnetic field. Nevertheless, active modulation of the surface charge on spacecraft could still enable effective orbital maneuvering. Recently, a series of applications of Lorentz spacecraft in space missions has been proposed, including spacecraft hovering [2], rendezvous [6], formation flying [1, 7–11], atmospheric drag compensation [3], orbital inclination change [12], planetary capture and escape [13–15], and so on. The necessary charging level varies with different space missions. Generally, the absolute orbital maneuvering, such as drag compensation and orbital inclination change, requires higher charging level than relative orbital maneuvering, such as spacecraft hovering and formation flying.

Spacecraft are naturally charged due to the surrounding plasma environment, and the natural charging level may reach to 10^{-8} C/kg, which is insufficient to perturb the orbit in a significant way [16]. Generally, in typical low Earth orbits (LEOs), a specific charge (i.e., charge-to-mass ratio) at least on the order of

© Springer Science+Business Media Singapore 2017

Y. Yan et al., *Dynamics and Control of Lorentz-Augmented Spacecraft Relative Motion*, DOI 10.1007/978-981-10-2603-4_1

10^{-5} C/kg is required to perform efficient orbital maneuvering [12]. It is generally accepted that a specific charge on the order of 10^{-3} to 10^{-2} C/kg is near-term feasible with concentrated research [14]. This charging level is sufficient for the relative orbital control, but is insufficient for absolute orbital control in LEOs. Also, given that the Lorentz force is determined by the local magnetic field and the vehicle's velocity relative to the magnetic field, a Lorentz spacecraft is thus more effective and efficient in LEOs where the magnetic field is intenser and the spacecraft travels faster than high Earth orbits. Moreover, the Lorentz spacecraft is ineffective in geosynchronous orbits (GEOs) due to the lack of relative velocity in such altitudes.

As compared to the propellant means of inter-spacecraft Coulomb force and solar radiation pressure, the geomagnetic Lorentz force is more favorable and advantageous in LEOs. A Coulomb spacecraft system consists of at least two charged spacecraft, of which the translational and rotational motion of the system is controlled by the inter-spacecraft Coulomb force. The Coulomb force is determined by the relative distance between spacecraft. If the relative distance is two times larger than the Debye shielding length, then the resulting Coulomb force will be negligibly small [17]. The Debye length varies with different orbital altitudes. It is on the order of several centimeters in LEOs, and reaches to the order of several kilometers in GEOs [17]. Thus, a Coulomb spacecraft system is much more applicable in high Earth orbits, and is hardly feasible in LEOs. Besides, the perturbation acceleration resulting from the solar radiation pressure is about on the order of 10^{-8} m/s^2 in LEOs [5], which is also too small to be applied in LEOs.

In view of these facts, this book concentrates on the applications of Lorentz spacecraft in relative orbital control in LEOs, including Lorentz-augmented spacecraft hovering, rendezvous, and formation flying.

1.2 Review

A brief review of the recent developments of Lorentz spacecraft is presented in this section. Also, the recent researches on the dynamics and control of spacecraft relative motion, including spacecraft hovering, rendezvous, and formation flying have been briefly reviewed.

1.2.1 Lorentz Spacecraft

The innovative concept of Lorentz spacecraft was first proposed by Peck [3], and the orbit propelled by the Lorentz force is termed as Lorentz-augmented orbit (LAO) [18]. The main idea of the LAO is inspired by the classical natural Lorentz force phenomenon in space, such as the motion of the charged dust in the planetary magnetic field [8]. Schaffer and Burns [19, 20] presented a detailed analysis on the orbital mechanics of the charged dust influenced by the Jupiter's gravitational and magnetic fields. Also, by using the dynamical model of LAO, the orbital evolution rules of circumplanetary dust by resonant charged variations have been revealed [21]. Thus, motivated by the concept of natural LAO where the dust or spacecraft is generally passively charged in the plasma environment, the concept of actively using the Lorentz force as electromagnetic propulsion for orbital maneuvering has received wide research interests.

Streetman and Peck [22] analyzed the effect of Lorentz force on the orbital elements, based on which new types of Earth-synchronous orbits in LEOs have been proposed. Meanwhile, the effect of Lorentz force on the gravity-assist maneuvers has also been examined [23]. As compared to the traditional gravity-assist maneuvers, the Lorentz-augmented one allows more flexible exit characteristics and timing of a fly. Furthermore, Streetman and Peck designed a general bang-bang control law for LAO [18].

Atchison and Peck [13] examined the feasibility of using the Lorentz force as a means of planetary capture or escape, and proposed a general design of a millimeter-scale Lorentz spacecraft [24]. Also, Atchison et al. [25] analyzed the flyby anomaly during low-altitude gravity-assist maneuvers around Earth, and indicated that the Lorentz force is unlikely to cause the flyby anomaly singly.

Gangestad et al. [2] proposed a control scheme for Lorentz-augmented station-keeping around small asteroids in space, and successfully applied it to the exploration of Enceladus, a moon of Saturn. Furthermore, they developed Lagrange's planetary equations for LAO, and presented analytical expressions that describe planetary capture propelled by the Lorentz force [14, 15].

Pollock et al. [12] studied the problem of orbital inclination change by the geomagnetic Lorentz force, and analyzed the enhanced response converge capability of a Lorentz spacecraft [26].

Besides the aforementioned researches mainly focused on absolute Lorentz-augmented orbital control, researches on relative Lorentz-augmented orbital control have also been conducted. The dynamical model of Lorentz-augmented relative motion is the fundament of the analysis on dynamics and control of Lorentz spacecraft relative motion. Hill–Clohessy–Wiltshire (HCW) [27] and Tschauner–Hempel (TH) [28] equations are classical models that characterize the orbital motion of an uncharged spacecraft relative to a circular and elliptic reference orbit, respectively. By modeling the Earth's magnetic field as a tilted dipole corotating with Earth, Pollock et al. [6] developed a dynamical model for Lorentz spacecraft relative motion about a circular reference orbit via incorporating the Lorentz acceleration into HCW equations. Then, approximate analytical solutions are derived for the following three cases: (1) equatorial circular reference orbit with nontilted dipole; (2) equatorial circular reference orbit with titled dipole; (3) inclined circular reference orbit with nontilted dipole. Similarly, based on the assumption of a nontilted magnetic dipole, Yamakawa et al. [29] developed a dynamical model for Lorentz spacecraft relative motion about an elliptic reference orbit via including the Lorentz acceleration into TH equations. Based on this model, they analyzed the stability of Lorentz-augmented relative orbital motion. As can be seen, present models do not include the orbital inclination and dipole tilt angle simultaneously. Actually, the Earth's magnetic dipole is tilted by 11.3° with respect to the Earth's rotation axis, a tilt angle not small enough to be neglected. Thus, when applying these models in Earth orbits, model errors are inevitably introduced.

By using these relative dynamical models, several applications have been proposed, such as Lorentz-augmented spacecraft formation flying (SFF), rendezvous and so on. Peck et al. [7] first proposed the concept of Lorentz-augmented SFF, and designed a reconfiguration scheme for triangular formation in equatorial plane. By including the Lorentz acceleration in the Gauss variational equations, Peng and Gao [8] designed a J_2-invariant formation augmented by the Lorentz force and proposed approximated analytical strategies on controlling the center and size of formations in circular reference orbits. Based on the assumption of a nontilted magnetic dipole, Tsujii et al. [1] studied the problem of relative orbit transfer propelled by the Lorentz force, and designed periodic relative orbits between Lorentz spacecraft with constant specific charge. Furthermore, Sobiesiak and Damaren [9, 10] analyzed the controllability of the dynamical system of Lorentz-augmented relative orbital motion, and concluded that the relative states between spacecraft are not fully controllable by using the Lorentz force only. Thus, other kinds of propulsion are necessary to render the system fully controllable. Based on this conclusion, they proposed optimal controllers for formation reconfiguration by using the hybrid propulsion consisting of the impulse and Lorentz force [11]. As to the application of Lorentz-augmented rendezvous, Pollock et al. [6] designed a rendezvous strategy in inclined circular orbits with nontilted dipole, and Yamakawa et al. [29] proposed another one in an equatorial elliptic orbit with nontilted dipole. As can be seen, both strategies are constrained in the nontilted dipole model, which is less representative of the Earth's magnetic field than a tilted one.

Furthermore, it is notable that the Lorentz force could also be used for spacecraft attitude control, besides the orbital control elaborated above. Yamakawa et al. [30] studied the attitude dynamics of a pendulum-shape charged spacecraft. In this work, two charged particles are connected by a rigid rod, and then the control torque resulting from the Lorentz force acting on both sides of the rod can be used for spacecraft attitude control. Also, Abdel-Aziz [31, 32] investigated the attitude dynamics and control of spacecraft using the Lorentz force and analyzed the stability of such attitude system. Furthermore, Giri and Sinha [33] investigated the attitude control of Earth-pointing satellite propelled by the geomagnetic Lorentz force via perturbation theory and averaging theory.

1.2.2 Spacecraft Relative Motion

Typical space missions associated with spacecraft relative motion include SFF, spacecraft rendezvous, and spacecraft hovering.

SFF refers to a cluster of smaller and less-expensive spacecraft flying in close proximity. It distributes the functionality of a traditional monolithic spacecraft to a few smaller ones [34]. Also, the spacecraft involved in a formation can reposition themselves to adapt to different space missions [35]. Thus, the advantages of SFF include enhanced flexibility and reliability, decreased risk and cost, etc. [36]. Given that detailed reviews of SFF haven been reported in the literature, reviews on SFF are not presented here for brevity, and the readers are referred to [37, 38] for details.

Spacecraft rendezvous refers to that two spacecraft reach the same position with zero relative velocity at the given final time. These two spacecraft are referred to as the chaser and target spacecraft, respectively. Generally, the target spacecraft is flying in a stable orbit, and no or little maneuvers are acted on the target. Contrarily, a series of maneuvers are acted on the chaser to achieve rendezvous with the target. Similarly, detailed reviews of spacecraft rendezvous have been reported in [39], and are thus not stated here again for brevity.

Spacecraft hovering denotes an equilibrium state in which the chaser maintains a constant relative position with respect to the target in space [40]. To achieve hovering, the chaser needs to thrust continuously to cancel the acceleration relative to the target, thus inducing an equilibrium state at the desired position [41]. Most hovering orbits are therefore non-Keplerian ones [42]. Compared to SFF or rendezvous, fewer reviews on hovering have been reported. Thus, a brief review on spacecraft hovering is presented as follows.

Hovering orbits hold promise in practical space missions. For example, hovering approach is preferable and advantageous in explorations of small asteroids or bodies in space, because it allows more high-resolution observation and measurements [41]. Hovering around the GEO could create another GEO at an orbital altitude other than the fixed 35,786 km [43]. In this way, this kind of precious orbital resource could be expanded. Furthermore, because the chaser and target are relatively static in a hovering configuration, the hovering approach allows more reliable proximity operations [44]. For example, for future fractionated spacecraft, the capability of hovering enables more reliable and efficient wireless energy transfer with less energy loss and decreased operation complexities [44].

Research on hovering can mainly be divided into two categories. One deals with hovering around small asteroids in space, the other deals with hovering around satellites in Earth orbits. For the former category, Lu and Love [45] designed a gravitational tractor that hovers above an Earth-threatening asteroid. This hovering spacecraft uses the gravity as a towline to alter the trajectory of the asteroid, thus reducing the threat of collision with Earth. Sawai et al. [41] proposed a closed-loop hovering strategy by using the altimetry measurements. Broschart and Scheeres [46, 47] performed a case study of hovering around an asteroid and analyzed the boundedness of spacecraft hovering under dead-band control. Also, Lee et al. [48, 49] proposed both asymptotic and finite-time control schemes for hovering over an asteroid.

As to the hovering around satellites in Earth orbits, Yan [42] derived open-loop control laws for hovering in circular orbits by using a geometry method. Then, this work is further extended to hovering in elliptic orbit by Zhu and Yan [50] via a similar method. Also, Zhang et al. [51] conducted analysis on the characteristic of hovering in two-body elliptic orbits via another dynamics method, and similar studies were performed by Dang et al. [52] in a J_2-perturted environment. Besides the aforementioned open-loop control laws for hovering, closed-loop ones have also been proposed via different control methods, such as linear quadratic regulator (LQR), sliding mode control (SMC), robust control, and so on. For example, in combination with frozen parameter method, Zhang et al. [53] designed a LQR for hovering in elliptic orbit. Also, another sliding mode controller was proposed by Wang et al. [43] to maintain the hovering position in the presence of external disturbances and system uncertainties. Zhou et al. [54, 55] designed both state and output feedback robust controllers for hovering. Furthermore, to deal with the case with the loss of the radial or in-track thrust, Huang et al. [56] investigated the feasibility of underactuated hovering and proposed corresponding underactuated controllers.

Notably, in all of the aforementioned hovering control schemes, whether the open-loop or closed-loop ones, the control acceleration necessary for hovering is assumed to be provided by traditional thrusters using chemical fuels. However, hovering can be fuel-consuming, especially in LEOs. For example, to hover radially 5 km above a target flying in a circular equatorial orbit with an orbital altitude of 500 km in a single day, it necessitates a velocity increment consumption of 1.586 km/s [44]. Such a considerable expenditure constraints the long-term hovering in LEOs via traditional thrusters [44]. Thus, novel means of propellantless propulsion will be preferable for long-term hovering in LEOs. As aforementioned, due to the considerable Debye shielding in LEOs, the Coulomb force can hardly be applied in LEOs. Thus, the hovering approach augmented by the inter-spacecraft Coulomb spacecraft, which was developed by Zhang et al. [17] for hovering around GEOs, can hardly be used in LEOs, too. Then, considering the enhanced efficiency of Lorentz spacecraft in LEOs, the geomagnetic Lorentz force will be a promising means of propulsion for long-term hovering in LEOs. Detailed analysis will be elaborated in this book.

Furthermore, besides detailed analysis on the dynamics and control of Lorentz-augmented spacecraft hovering, studies on Lorentz-augmented SFF and rendezvous have also been conducted in this book. A systematic description of the organization of this book will be given in the following section.

1.3 Outline of the Book

The remainder of this book consists of five chapters.

Chapter 2 develops the dynamical models of Lorentz-augmented absolute and relative orbital motion in both two-body and J_2-perturbed environment. Different from previous models, the models in this book explicitly include the orbital inclination and dipole tilt angle simultaneously, which is more representative of the Earth's magnetic field and is expected to present enhanced accuracy. Also, approximated analytical solutions are derived for relative motion about a circular inclined reference orbit with tilted magnetic dipole. Meanwhile, the effect of J_2 perturbation on Lorentz-augmented relative orbital motion is numerically studied.

Chapter 3 designs extended Kalman filter (EKF) and unscented Kalman filter (UKF) for six-degree-of-freedom (6DOF) relative navigation of Lorentz spacecraft. The estimated states consist of relative position, velocity, attitude quaternion, and angular velocity between spacecraft. Both filters are developed based on newly developed dynamical models and the observation system consisting of the line-of-sight observations and gyro measurements. Also, performances of the filters are numerically compared.

Chapter 4 elaborates the dynamics and control of hovering augmented by the geomagnetic Lorentz force. Both circular and elliptic references orbits are considered. For circular reference orbits, the hovering configurations that could achieve propellantless hovering are first derived, and the corresponding required specific charges of Lorentz spacecraft are also presented. For other hovering configurations that necessitate the hybrid propulsion consisting of the Lorentz force and thruster-generated control force, an energy-optimal distribution law of these two kinds of propulsion is derived via a Lagrangian method. For elliptic reference orbits, similar optimal distribution law of the two kinds of propulsion is also proposed. Considering the external perturbations and system uncertainties that may perturb the hovering position, closed-loop controllers are designed for both cases, too. Furthermore, the effect of J_2 perturbation on Lorentz-augmented spacecraft hovering is also studied.

Chapter 5 discusses the dynamics and control of Lorentz-augmented SFF in circular or elliptic orbits. A brief review of formation conditions in circular or elliptic orbits is first presented. By explicitly introducing these conditions, the optimal control problem of Lorentz-augmented formation establishment or reconfiguration is formulated as a constrained trajectory optimization problem (TOP). The TOP is further transcribed into a nonlinear programming (NLP) by using the Gauss pseudospectral method (GPM), and the resulting NLP is then solved by appropriate numerical methods. Notably, both fixed and free final condition constraints are considered. For the fixed one, the terminal relative states are set as priori determined fixed points. Differently, for the free one, the terminal relative states are regarded as optimization variables, which are not necessarily determined in advance. Comparisons are then made between these two cases. Likewise, closed-loop controllers are also proposed for Lorentz-augmented SFF in circular or elliptic orbits, aiming to ensure optimal trajectory tracking in the presence of initial offsets and external disturbances.

Chapter 6 investigates the dynamics and control of Lorentz-augmented rendezvous in circular or elliptic orbits. Similarly, the optimal control problem of Lorentz-augmented spacecraft rendezvous is formulated as a general constrained TOP, which is further transcribed into NLP by GPM and solved by numerical optimization approaches. In generating the optimal rendezvous trajectory, cases with both fixed and free final time are considered, and comparisons are made between the optimal results of these two cases. The optimal Lorentz-augmented rendezvous strategies proposed in this book avoid the constraints imposed on the initial relative states, reference orbits, and charging time in previous rendezvous strategies. Also, the effect of J_2 perturbation on the optimal Lorentz-augmented rendezvous trajectory is numerically studied.

References

1. Tsujii S, Bando M, Yamakawa H (2013) Spacecraft formation flying dynamics and control using the geomagnetic Lorentz force. J Guid Control Dyn 36:136–148
2. Gangestad JW, Pollock GE, Longuski JM (2009) Propellantless stationkeeping at Enceladus via the electromagnetic Lorentz force. J Guid Control Dyn 32:1466–1475
3. Peck MA (2005) Prospects and challenges for Lorentz-augmented orbits. In: AIAA guidance, navigation, and control conference and exhibit, San Francisco, 15–18 Aug 2005
4. Schaub H, Parker GG, King LB (2003) Challenges and prospect of Coulomb formations. In: AAS John L. Junkins astrodynamics symposium, Texas, 23–24 May 2003
5. Gong S, Li J (2015) Dynamics and control of solar sail spacecraft. Tsinghua, Beijing
6. Pollock GE, Gangestad JW, Longuski JM (2011) Analytical solutions for the relative motion of spacecraft subject to Lorentz-force perturbations. Acta Astronaut 68:204–217
7. Peck MA, Streetman B, Saaj CM et al (2007) Spacecraft formation flying using Lorentz force. J Br Interplanet Soc 60:263–267
8. Peng C, Gao Y (2012) Lorentz-force-perturbed orbits with application to J_2-invariant formation. Acta Astronaut 77:12–28
9. Sobiesiak LA, Damaren CJ (2015) Optimal continuous/impulsive control for Lorentz-augmented spacecraft formations. J Guid Control Dyn 38:151–156
10. Sobiesiak LA, Damaren CJ (2015) Controllability of Lorentz-augmented spacecraft formations. J Guid Control Dyn 38:2188–2195
11. Sobiesiak LA, Damaren CJ (2016) Lorentz-augmented spacecraft formation reconfiguration. IEEE Trans Control Syst Technol 24:514–524
12. Pollock GE, Gangestad JW, Longuski JM (2010) Inclination change in low-Earth orbit via the geomagnetic Lorentz force. J Guid Control Dyn 33:1387–1395
13. Atchison JA, Peck MA (2009) Lorentz-augmented Jovian orbit insertion. J Guid Control Dyn 32:418–423
14. Gangestad JW, Pollock GE, Longuski JM (2010) Lagrange's planetary equations for the motion of electrostatically charged spacecraft. Celest Mech Dyn Astron 108:125–145
15. Gangestad JW, Pollock GE, Longuski JM (2011) Analytical expressions that characterize propellantless capture with electrostatically charged spacecraft. J Guid Control Dyn 34: 247–258
16. Vokrouhlický D (1989) The geomagnetic effects on the motion of an electrically charged artificial satellite. Celest Mech Dyn Astron 46:85–104
17. Zhang H, Shi P, Li B et al (2012) Hovering orbit using inter-spacecraft Coulomb forces. J Astronaut 33:68–75
18. Streetman B, Peck MA (2010) General bang-bang control method for Lorentz augmented orbits. J Spacecr Rockets 47:484–492
19. Schaffer L, Burns JA (1987) The dynamics of weakly charged dust: motion through Jupiter's gravitational and magnetic fields. J Geophys Res 92:2264–2280
20. Schaffer L, Burns JA (1994) Charged dust in planetary magnetospheres: Hamiltonian dynamics and numerical simulations for highly charged grains. J Geophys Res 99: 17, 211–217, 223
21. Burns JA, Schaffer L (1989) Orbital evolution of circumplanetary dust by resonant charged variations. Nature 337:340–343

22. Streetman B, Peck MA (2007) New synchronous orbits using the geomagnetic Lorentz force. J Guid Control Dyn 30:1677–1690
23. Streetman B, Peck MA (2009) Gravity-assist maneuvers augmented by the Lorentz force. J Guid Control Dyn 32:1639–1647
24. Atchison JA, Peck MA (2007) A millimeter-scale Lorentz-propelled spacecraft. In: AIAA guidance, navigation, and control conference and exhibit, Hilton Head, 20–23 Aug 2007
25. Atchison JA, Peck MA, Streetman BJ (2010) Lorentz accelerations in the Earth flyby anomaly. J Guid Control Dyn 33:1115–1122
26. Pollock GE, Gangestad JW, Longuski JM (2008) Responsive coverage using propellantless satellites. In: 6th responsive space conference, Los Angeles, 28 April–1 May 2008
27. Clohessy W, Wiltshire R (1960) Terminal guidance system for satellite rendezvous. J Aerosp Eng 27:653–658
28. Tschauner J, Hempel P (1965) Rendezvous with a target in an elliptical orbit. Acta Astronaut 11:104–109
29. Yamakawa H, Bando M, Yano K et al (2010) Spacecraft relative dynamics under the influence of geomagnetic Lorentz force. In: AIAA/AAS astrodynamics specialist conference, Toronto, 2–5 Aug 2010
30. Yamakawa H, Hachiyama S, Bando M (2012) Attitude dynamics of a pendulum-shaped charged satellite. Acta Astronaut 70:77–84
31. Abdel-Aziz YA (2007) Attitude stabilization of a rigid spacecraft in the geomagnetic field. Adv Space Res 40:18–24
32. Abdel-Aziz YA, Shoaib M (2014) Numerical analysis of attitude stability of a charged spacecraft in the pitch-roll-yaw directions. Int J Aeronaut Space Sci 15:82–90
33. Giri DK, Sinha M (2014) Magnetocoulombic attitude control of Earth-pointing satellites. J Guid Control Dyn 37:1946–1960
34. Inalhan G, Tillerson M, How JP (2002) Relative dynamics and control of spacecraft formations in eccentric orbits. J Guid Control Dyn 25:48–59
35. Sabol C, Burns R, McLaughlin CA (2001) Spacecraft formation design and evolution. J Spacecr Rockets 38:270–278
36. Lee D, Sanyal AK, Butcher EA (2015) Asymptotic tracking control for spacecraft formation flying with decentralized collision avoidance. J Guid Control Dyn 38:587–600
37. Kristiansen R, Nicklasson RJ (2009) Spacecraft formation flying: a review and new results on state feedback control. Acta Astronaut 65:1537–1552
38. Alfriend KT, Vadali SR, Gurfil P et al (2010) Spacecraft formation flying. Butterworth-Heinemann, Oxford
39. Luo Y, Zhang J, Tang G (2014) Survey of orbital dynamics and control of spacecraft rendezvous. Chin J Aeronaut 27:1–11
40. Huang X, Yan Y, Zhou Y et al (2015) Output feedback sliding mode control of Lorentz-augmented spacecraft hovering using neural networks. Proc Inst Mech Eng Part I J Syst Control Eng 229:939–948
41. Sawai S, Scheeres DJ, Broschart SB (2002) Control of hovering spacecraft using altimetry. J Guid Control Dyn 25:786–795
42. Yan Y (2009) Study of hovering method at any selected position to space target. Chin Space Technol Sci 29:1–5
43. Wang G, Zheng W, Meng Y et al (2011) Research on hovering control scheme to non-circular orbit. Sci China Technol Sci 54:2974–2980
44. Huang X, Yan Y, Zhou Y (2014) Dynamics and control of spacecraft hovering using the geomagnetic Lorentz force. Adv Space Res 53:518–531
45. Lu ET, Love SG (2005) A gravitational tractor for towing asteroids. Nature 438:177–178
46. Broschart SB, Scheeres DJ (2005) Control of hovering spacecraft near small bodies: application to asteroid 25143 Itokawa. J Guid Control Dyn 28:343–354
47. Broschart SB, Scheeres DJ (2007) Boundedness of spacecraft hovering under dead-band control in time-invariant systems. J Guid Control Dyn 30:601–610

48. Lee D, Sanyal AK, Butcher EA et al (2014) Almost global asymptotic tracking control for spacecraft body-fixed hovering over an asteroid. Aerosp Sci Technol 38:105–115
49. Lee D, Sanyal AK, Butcher EA et al (2015) Finite-time control for spacecraft body-fixed hovering over an asteroid. IEEE Trans Aerosp Electron Syst 51:506–520
50. Zhu Y, Yan Y (2010) Hovering method at any selected position over space target on elliptical orbit. Chin Space Technol Sci 30(17–23):48
51. Zhang J, Zhao S, Yang Y (2013) Characteristic analysis for elliptical orbit hovering based on relative dynamics. IEEE Trans Aerosp Electron Syst 49:2742–2750
52. Dang Z, Wang Z, Zhang Y (2014) Modeling and analysis of relative hovering control for spacecraft. J Guid Control Dyn 37:1091–1102
53. Zhang J, Zhao S, Zhang Y (2015) Hovering control scheme to elliptical orbit via frozen parameter. Adv Space Res 55:334–342
54. Zhou Y, Yan Y, Huang X (2015) Multi-objective robust controller design for spacecraft hovering around elliptical orbital target. Proc Inst Mech Eng Part I J Syst Control Eng 229:356–371
55. Zhou Y, Yan Y, Huang X et al (2015) Multi-objective and reliable output feedback control for spacecraft hovering. Proc Inst Mech Eng Part G J Aerosp Eng 229:1798–1812
56. Huang X, Yan Y, Zhou Y (2016) Nonlinear control of underactuated spacecraft hovering. J Guid Control Dyn 39:684–693

Chapter 2
Dynamical Model of Lorentz-Augmented Orbital Motion

Dynamics analysis and controller design of Lorentz-augmented space missions rely highly on the precise modeling of Lorentz-augmented orbital motion. Current models for Lorentz-augmented absolute or relative orbital motion, especially the relative ones, are mostly developed based on the assumption of a nontilted magnetic dipole. However, the Earth's magnetic dipole is tilted by about 11.3° with respect to the rotation axis of Earth, which is not small enough to be neglected. In view of this fact, this chapter develops dynamical models for Lorentz-augmented absolute and relative orbital motion that explicitly include the dipole tilt angle in both two-body and J_2-perturbed environment.

2.1 Model of Absolute Orbital Motion

2.1.1 Geomagnetic Field

The potential function of the geomagnetic field can be expressed as [1]

$$V = R_E \sum_{n=1}^{\infty} \sum_{m=0}^{n} \left\{ P_n^m(\cos\theta)[g_n^m(t)\cos(m\lambda) + h_n^m(t)\sin(m\lambda)](R_E/r)^{n+1} \right\} \quad (2.1)$$

where $R_E = 6371.2$ km is the magnetic reference spherical radius of Earth, and r is the orbital radius. θ and λ refer to the geocentric colatitude and east longitude, respectively. $g_n^m(t)$ and $h_n^m(t)$ are time-dependent Gauss coefficients, which can be derived by referring to the International Geomagnetic Reference Field (IGRF) [1].

$P_n^m(\cos\theta)$ refers to the Schmidt quasi-normalized Legendre function of degree n and order m, given by [2]

© Springer Science+Business Media Singapore 2017

Y. Yan et al., *Dynamics and Control of Lorentz-Augmented Spacecraft Relative Motion*, DOI 10.1007/978-981-10-2603-4_2

$$P_n^m(\cos\theta) = \frac{1}{2^n \cdot n!}\left[\frac{\varepsilon_m(n-m)!(1-\cos^2\theta)^m}{(n+m)!}\right]^{1/2}\frac{\mathrm{d}^{n+m}}{\mathrm{d}(\cos\theta)^{n+m}}(\cos^2\theta-1)^n \tag{2.2}$$

where

$$\varepsilon_m = \begin{cases} 1, & m=0 \\ 2, & m\geq 1 \end{cases}, n\geq m\geq 0 \tag{2.3}$$

The magnetic field $\boldsymbol{B}$ can then be derived as the negative gradient of the potential, given by

$$\boldsymbol{B} = -\nabla V \tag{2.4}$$

As depicted in Fig. 2.1, $O_E X_I Y_I Z_I$ is an Earth-centered inertial (ECI) frame, and $O_E X_E Y_E Z_E$ is another Earth-fixed frame that is corotating with Earth. O_E is the center of Earth. O_L refers to the center of mass (c.m.) of the Lorentz spacecraft, where the local geocentric longitude and colatitude are λ and θ, respectively. Define B_θ, B_λ, and B_r as the components of the local magnetic field $\boldsymbol{B}$ in the northward, eastward, and radially inward directions, respectively. By using Eq. (2.4), it can be derived that [2]

$$\begin{aligned} B_\theta &= \frac{1}{r}\frac{\partial V}{\partial\theta} \\ &= \sum_{n=1}^{N}(R_E/r)^{n+2}\sum_{m=0}^{n}\left[g_n^m\cos(m\lambda)+h_n^m\sin(m\lambda)\right]\frac{\partial P_n^m(\cos\theta)}{\partial\theta} \end{aligned} \tag{2.5}$$

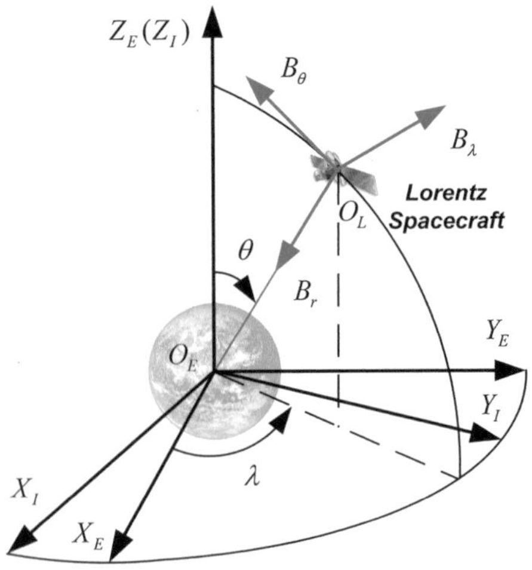

Fig. 2.1 Illustration of the decomposition of the geomagnetic field

$$
\begin{aligned}
B_\lambda &= -\frac{1}{r \sin\theta} \frac{\partial V}{\partial \lambda_L} \\
&= \sum_{n=1}^{N} (R_E/r)^{n+2} \sum_{m=0}^{n} m\left[g_n^m \sin(m\lambda) - h_n^m \cos(m\lambda)\right] \frac{P_n^m(\cos\theta)}{\sin\theta}
\end{aligned} \tag{2.6}
$$

$$
\begin{aligned}
B_r &= \frac{\partial V}{\partial r} \\
&= -\sum_{n=1}^{N} (R_E/r)^{n+2} (n+1) \sum_{m=0}^{n} \left[g_n^m \cos(m\lambda) + h_n^m \sin(m\lambda)\right] P_n^m(\cos\theta)
\end{aligned} \tag{2.7}
$$

As shown Fig. 2.2, B_F is the total intensity, B_H is the horizontal intensity, D is the declination angle, and I is the inclination angle. Based on the definitions in Fig. 2.2, these variables can be calculated via [1]

$$
\begin{aligned}
B_F &= \sqrt{B_\theta^2 + B_\lambda^2 + B_r^2},\ B_H = \sqrt{B_\theta^2 + B_\lambda^2} \\
D &= \arctan(B_\lambda/B_\theta),\ I = \arctan(B_r/B_H)
\end{aligned} \tag{2.8}
$$

According to the 11th generation of IGRF (IGRF-11), the magnetic field information at the orbital altitude of 566.9 km in 2010 is shown from Figs. 2.3, 2.4 and 2.5.

If the series in Eq. (2.1) is truncated at $n = 1$ that only g_1^0, g_1^1, and h_1^1 remain, then the geomagnetic field reduces to a tilted dipole model. Each of the three Gauss coefficients corresponds to a dipole located at the center of Earth. Notably, these three dipoles are perpendicular with respect to each other. Thus, the composed dipole is still located at the center of Earth, but is tilted by an angle with respect to the rotation axis of Earth.

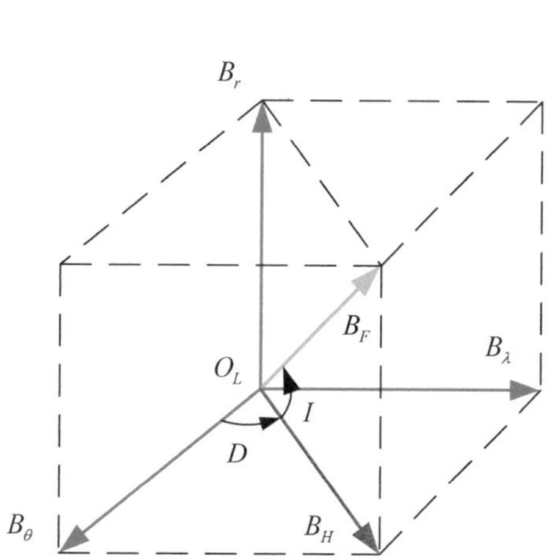

Fig. 2.2 Definitions of the related variables

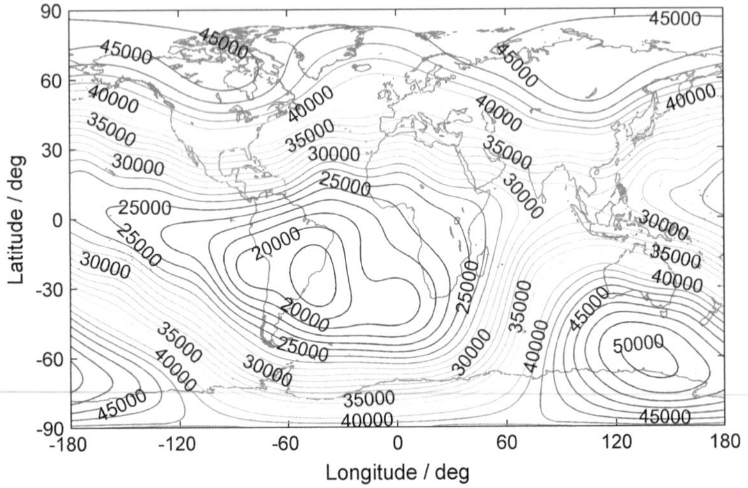

Fig. 2.3 Equivalence graph of the total intensity (unit: nT)

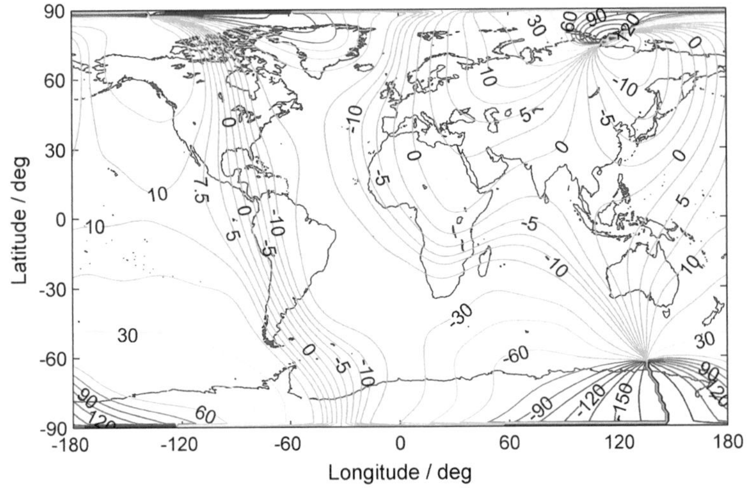

Fig. 2.4 Equivalence graph of the declination angle (unit: deg)

By modeling the geomagnetic field as a tilted dipole corotating with Earth, the magnetic vector potential can be represented as [3]

$$\mathbf{A} = \frac{B_0}{r^2}(\hat{\boldsymbol{n}} \times \hat{\boldsymbol{r}}) \tag{2.9}$$

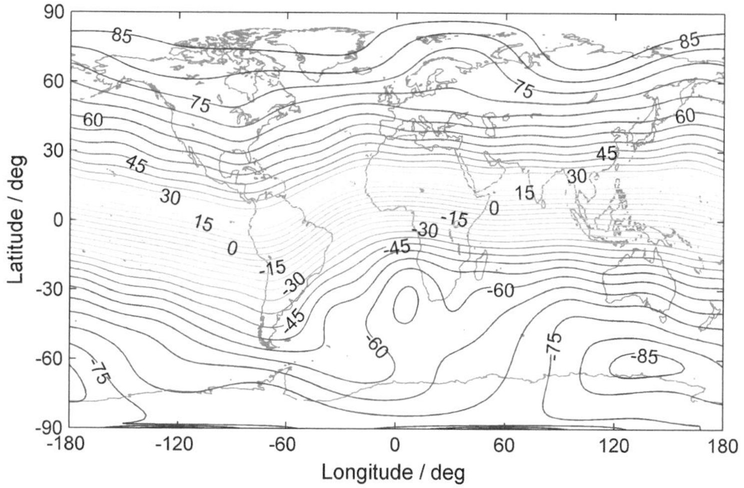

Fig. 2.5 Equivalence graph of the inclination angle (unit: deg)

where $B_0 = 8 \times 10^{15}\mathrm{T\ m}^3$ is the magnetic dipole moment of Earth, and r is the orbital radius. The superscript ^ refers to a unit vector in that direction. For example, $\hat{\boldsymbol{n}}$ is the unit vector in the direction of the magnetic dipole moment and $\hat{\boldsymbol{r}}$ is the unit orbital radius vector.

The magnetic field located at $\boldsymbol{r}$ can be then calculated via [3]

$$\boldsymbol{B} = \nabla \times \boldsymbol{A} = \frac{B_0}{r^3}[3(\hat{\boldsymbol{n}} \cdot \hat{\boldsymbol{r}})\hat{\boldsymbol{r}} - \hat{\boldsymbol{n}}] \tag{2.10}$$

2.1.2 *Two-Body Model*

As shown in Fig. 2.6, $O_E X_I Y_I Z_I$ is the ECI frame, and the angles α and Ω_M define the orientation of the magnetic dipole in the ECI frame, where α is the dipole tilt angle with respect to the rotation axis of Earth, Ω_M is the inertial rotation angle of the magnetic dipole, given by

$$\Omega_M = \omega_E t + \Omega_0 \tag{2.11}$$

where ω_E is the rotation rate of Earth, t denotes the time, and Ω_0 is the initial phase angle of the magnetic dipole.

Based on the above definitions, the unit vector $\hat{\boldsymbol{n}}$ can be expressed in the ECI frame as

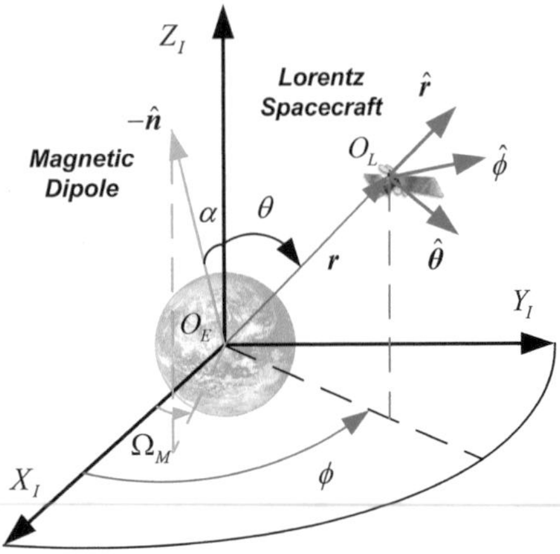

Fig. 2.6 Definitions of the spherical coordinate frame and the dipole orientation. Reprinted from Ref. [11], Copyright 2015, with permission from SAGE

$$\hat{\boldsymbol{n}} = -\cos\Omega_M \sin\alpha\hat{\boldsymbol{X}}_I - \sin\Omega_M \sin\alpha\hat{\boldsymbol{Y}}_I - \cos\alpha\hat{\boldsymbol{Z}}_I \tag{2.12}$$

where $\hat{\boldsymbol{X}}_I$, $\hat{\boldsymbol{Y}}_I$, and $\hat{\boldsymbol{Z}}_I$ refer to the unit vector along the three axes of ECI frame, respectively.

In Fig. 2.6, O_L is the c.m. of the Lorentz spacecraft, the spherical inertial coordinates r, ϕ, and θ give the position of O_L in ECI frame. Thus, the orbital radius vector of the Lorentz spacecraft is given by

$$\boldsymbol{r} = r\cos\phi\,\sin\theta\hat{\boldsymbol{X}}_I + r\sin\phi\,\sin\theta\hat{\boldsymbol{Y}}_I + r\cos\theta\hat{\boldsymbol{Z}}_I \tag{2.13}$$

Also, the unit orbital radius vector $\hat{\boldsymbol{r}}$ is

$$\hat{\boldsymbol{r}} = \cos\phi\,\sin\theta\hat{\boldsymbol{X}}_I + \sin\phi\,\sin\theta\hat{\boldsymbol{Y}}_I + \cos\theta\hat{\boldsymbol{Z}}_I \tag{2.14}$$

Taking the time derivative of Eq. (2.13) yields the velocity of Lorentz spacecraft, given by

$$\boldsymbol{v} = \frac{\mathrm{d}\boldsymbol{r}}{\mathrm{d}t} = \begin{bmatrix} \dot{r}\cos\phi\,\sin\theta - r\dot{\phi}\,\sin\phi\,\sin\theta + r\dot{\theta}\,\cos\phi\,\cos\theta \\ \dot{r}\sin\phi\,\sin\theta + r\dot{\phi}\,\cos\phi\,\sin\theta + r\dot{\theta}\,\sin\phi\,\cos\theta \\ \dot{r}\cos\theta - r\dot{\theta}\,\sin\theta \end{bmatrix} \tag{2.15}$$

Then, the kinetic energy per unit mass of Lorentz spacecraft is derived as

$$T = \frac{1}{2}\boldsymbol{v}^{\mathrm{T}}\boldsymbol{v} = \frac{1}{2}(\dot{r}^2 + r^2\dot{\theta}^2 + r^2\dot{\phi}^2\sin^2\theta) \tag{2.16}$$

The velocity-dependent magnetic potential of unit mass of Lorentz spacecraft is [4]

$$U_L = -\lambda_L \boldsymbol{V}_{\text{rel}}^{\text{T}} \boldsymbol{A} \tag{2.17}$$

where $\lambda_L = q_L/m_L$ is the specific charge of Lorentz spacecraft (i.e., charge-to-mass ratio), with q_L and m_L being the charge and mass of Lorentz spacecraft, respectively. $\boldsymbol{V}_{\text{rel}}$ is the velocity of Lorentz spacecraft with respect to the local magnetic field, given by

$$\boldsymbol{V}_{\text{rel}} = \boldsymbol{v} - \boldsymbol{\omega}_E \times \boldsymbol{r} \tag{2.18}$$

where $\boldsymbol{\omega}_E = \begin{bmatrix} 0 & 0 & \omega_E \end{bmatrix}^{\text{T}}$ is the angular velocity vector of Earth in ECI frame.

In a two-body model, the gravitational potential per unit mass of Lorentz spacecraft is

$$U_g = -\frac{\mu}{r} \tag{2.19}$$

where μ is the gravitational parameter of Earth.

Therefore, the Lagrange function per unit mass of Lorentz spacecraft can be summarized as

$$L = T - U_g - U_L \tag{2.20}$$

Substitution of Eqs. (2.16), (2.17), and (2.19) into Eq. (2.20) yields that [3]

$$\begin{aligned} L = &\frac{1}{2}(\dot{r}^2 + \dot{\theta}^2 r^2 + \dot{\phi}^2 r^2 \sin^2\theta) + \frac{\mu}{r} + \frac{q_L B_0}{m_L r}\{\dot{\theta}\sin(\phi - \Omega_M)\sin\alpha \\ &+ (\dot{\phi} - \omega_E)\sin\theta[\cos\theta\cos(\phi - \Omega_M)\sin\alpha - \sin\theta\cos\alpha]\} \end{aligned} \tag{2.21}$$

By solving the Lagrangian equation

$$\frac{\mathrm{d}}{\mathrm{d}t}\left(\frac{\partial L}{\partial \dot{\boldsymbol{\chi}}}\right) - \frac{\partial L}{\partial \boldsymbol{\chi}} = 0 \tag{2.22}$$

where $\boldsymbol{\chi} = \begin{bmatrix} r & \phi & \theta \end{bmatrix}^{\text{T}}$ denotes the generalized coordinates, the dynamic equations of the two-body absolute orbital motion of Lorentz spacecraft can be obtained as [3]

$$\begin{aligned} \ddot{r} = &-\frac{q_L B_0}{m_L r^2}\{\dot{\theta}\sin(\phi - \Omega_M)\sin\alpha + (\dot{\phi} - \omega_E)\sin\theta[-\sin\theta\cos\alpha \\ &+ \cos\theta\cos(\phi - \Omega_M)\sin\alpha]\} + r\dot{\theta}^2 + r\dot{\phi}^2\sin^2\theta - \frac{\mu}{r^2} \end{aligned} \tag{2.23}$$

$$\begin{aligned} \ddot{\phi} = &-\frac{q_L B_0}{m_L r^4}[\cos\alpha(\dot{r} - 2r\dot{\theta}\cot\theta) - \cos(\phi - \Omega_M)\sin\alpha(\dot{r}\cot\theta + 2r\dot{\theta})] \\ &- 2\dot{\phi}\frac{\dot{r}}{r} - 2\dot{\phi}\dot{\theta}\cot\theta \end{aligned} \tag{2.24}$$

$$\begin{aligned}\ddot{\theta} = & \frac{q_L B_0}{m_L r^3}\{(\dot{\phi} - \omega_E)[(\cos 2\theta - 1)\cos(\phi - \Omega_M)\sin\alpha - \sin 2\theta \cos\alpha] \\ & + [\dot{r}\sin(\phi - \Omega_M)]/r\} - 2\dot{\theta}\frac{\dot{r}}{r} + \dot{\phi}^2 \sin\theta\cos\theta\end{aligned} \tag{2.25}$$

Notably, Eq. (2.24) rectifies the typos in Eq. (2.15) of Ref. [3].

2.1.3 J_2-Perturbed Model

A Lorentz spacecraft is more effective and efficient in low Earth orbits (LEOs) where the magnetic field is intenser and the spacecraft travels faster than high Earth orbits. Moreover, J_2 perturbation, arising from the oblateness and nonhomogeneity of Earth, is one of the most dominant disturbances in LEOs. Therefore, to capture the environment in LEOs more precisely, a dynamical model that characterizes the J_2-perturbed orbital motion of Lorentz spacecraft is developed as follows.

In consideration of the J_2 perturbation, the gravitational potential per unit mass of Lorentz spacecraft is revised as [5]

$$U_{g,J_2} = -\frac{\mu}{r}\left[1 - J_2(R_E/r)^2(3\cos^2\theta - 1)/2\right] \tag{2.26}$$

where is $J_2 = 1.0826 \times 10^{-3}$ the second zonal harmonic coefficient of Earth.

Correspondingly, the Lagrange function is revised as

$$L = T - U_{g,J_2} - U_L \tag{2.27}$$

where the kinetic energy and magnetic potential per unit mass are the same as those in Eqs. (2.16) and (2.17).

Similarly, by solving the Lagrangian equation, the dynamic equations of J_2-perturbed orbital motion of Lorentz spacecraft can be derived as

$$\begin{aligned}\ddot{r} = & -\frac{q_L B_0}{m_L r^2}\left\{\dot{\theta}\sin(\phi - \Omega_M)\sin\alpha + (\dot{\phi} - \omega_E)\sin\theta[\cos\theta\cos(\phi - \Omega_M)\sin\alpha\right. \\ & \left. - \sin\theta\cos\alpha]\right\} + r\dot{\theta}^2 + r\dot{\phi}^2\sin^2\theta - \frac{\mu}{r^2} + \frac{3}{2}\mu J_2\frac{R_E^2}{r^4}(3\cos^2\theta - 1)\end{aligned} \tag{2.28}$$

$$\begin{aligned}\ddot{\phi} = & -\frac{q_L B_0}{m_L r^4}\left[\cos\alpha(\dot{r} - 2r\dot{\theta}\cot\theta) - \cos(\phi - \Omega_M)\sin\alpha(\dot{r}\cot\theta + 2r\dot{\theta})\right] \\ & - 2\dot{\phi}\frac{\dot{r}}{r} - 2\dot{\phi}\dot{\theta}\cot\theta\end{aligned} \tag{2.29}$$

$$\ddot{\theta} = \frac{q_L B_0}{m_L r^3}\left\{(\dot{\phi} - \omega_E)[(\cos 2\theta - 1)\cos(\phi - \Omega_M)\sin\alpha - \sin 2\theta \cos\alpha]\right. \\ \left. + [\dot{r}\sin(\phi - \Omega_M)\sin\alpha]/r\right\} - 2\dot{\theta}\frac{\dot{r}}{r} + \dot{\phi}^2 \sin\theta \cos\theta + \frac{3}{2}\mu J_2 \frac{R_E^2}{r^5}\sin 2\theta \tag{2.30}$$

2.2 Model of Relative Orbital Motion

2.2.1 Two-Body Model

2.2.1.1 Equations of Relative Motion

Consider a chaser spacecraft and another target spacecraft flying in Earth orbits. The chaser is assumed to be a charged Lorentz spacecraft, but the target is uncharged. As shown in Fig. 2.7, $O_E X_I Y_I Z_I$ is the ECI frame. $O_T xyz$ is a local-vertical-local-horizontal (LVLH) frame located at the c.m. of the target, O_T, where x axis is along the radial direction, z axis is normal to the target's orbital plane, and y axis completes the right-handed Cartesian frame. $\boldsymbol{R}_L$ and $\boldsymbol{R}_T$ refer to the orbital radius vector of the chaser and the target, respectively. The relative position vector between the chaser and the target is thus $\boldsymbol{\rho} = \boldsymbol{R}_L - \boldsymbol{R}_T = [x \quad y \quad z]^{\mathrm{T}}.\ i_T,$

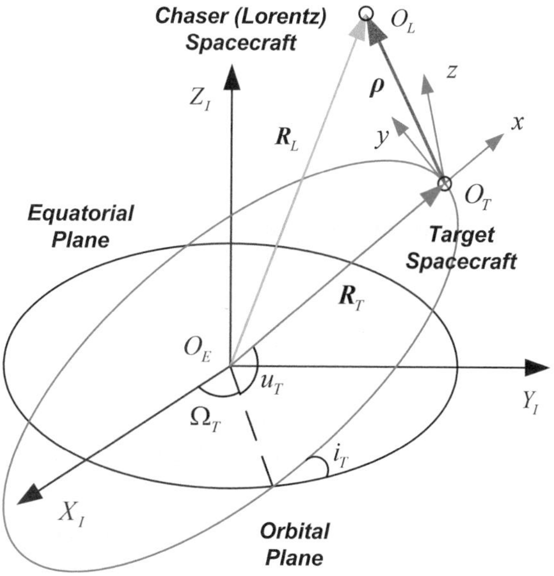

Fig. 2.7 Definitions of the coordinate frames. Reprinted from Ref. [11], Copyright 2015, with permission from SAGE

Ω_T, and u_T are, respectively, the inclination, right ascension of the ascending node, and argument of latitude of the target.

The equations of orbital motion of the Lorentz and the target spacecraft are, respectively, given by

$$\frac{\mathrm{d}^2 \boldsymbol{R}_L}{\mathrm{d}t^2} = -\frac{\mu}{R_L^3}\boldsymbol{R}_L + \boldsymbol{a}_L + \boldsymbol{a}_C \tag{2.31}$$

$$\frac{\mathrm{d}^2 \boldsymbol{R}_T}{\mathrm{d}t^2} = -\frac{\mu}{R_T^3}\boldsymbol{R}_T \tag{2.32}$$

where $\boldsymbol{a}_L = \begin{bmatrix} a_x & a_y & a_z \end{bmatrix}^{\mathrm{T}}$ and $\boldsymbol{a}_C = \begin{bmatrix} a_R & a_S & a_W \end{bmatrix}^{\mathrm{T}}$ refer to the Lorentz acceleration and control acceleration acting on the Lorentz spacecraft, respectively.

Given that $\boldsymbol{\rho} = \boldsymbol{R}_L - \boldsymbol{R}_T$, taking the second order time derivative of this equation yields that

$$\frac{\mathrm{d}^2 \boldsymbol{\rho}}{\mathrm{d}t^2} = \ddot{\boldsymbol{\rho}} + \dot{\boldsymbol{u}}_T \times (\dot{\boldsymbol{u}}_T \times \boldsymbol{\rho}) + 2\dot{\boldsymbol{u}}_T \times \dot{\boldsymbol{\rho}} + \ddot{\boldsymbol{u}}_T \times \boldsymbol{\rho} = \frac{\mathrm{d}^2 \boldsymbol{R}_L}{\mathrm{d}t^2} - \frac{\mathrm{d}^2 \boldsymbol{R}_T}{\mathrm{d}t^2} \tag{2.33}$$

where $\dot{\boldsymbol{u}}_T = \begin{bmatrix} 0 & 0 & \dot{u}_T \end{bmatrix}^{\mathrm{T}}$ and $\ddot{\boldsymbol{u}}_T = \begin{bmatrix} 0 & 0 & \ddot{u}_T \end{bmatrix}^{\mathrm{T}}$ refer to orbital angular velocity and acceleration of the target, respectively.

Substitution of Eqs. (2.31) and (2.32) into Eq. (2.33) yields the nonlinear equations of relative orbital motion of Lorentz spacecraft expressed in the LVLH frame, given by

$$\begin{aligned} \ddot{x} &= 2\dot{u}_T\dot{y} + \dot{u}_T^2 x + \ddot{u}_T y + n_T^2 R_T - n_L^2 (R_T + x) + a_x + a_R \\ \ddot{y} &= -2\dot{u}_T\dot{x} + \dot{u}_T^2 y - \ddot{u}_T x - n_L^2 y + a_y + a_S \\ \ddot{z} &= -n_L^2 z + a_z + a_W \end{aligned} \tag{2.34}$$

where $n_T = \sqrt{\mu/R_T^3}$ and $n_L = \sqrt{\mu/R_L^3}$. $R_L = [(R_T + x)^2 + y^2 + z^2]^{1/2}$ is the orbital radius of Lorentz spacecraft.

Provided that the relative distance between spacecraft is negligibly small compared to the orbital radius of either spacecraft (i.e., $||\boldsymbol{\rho}|| \ll R_L, R_T$), the above nonlinear equations can be further linearized. Based on this precondition, it holds that

$$\left(\frac{R_T}{R_L}\right)^3 = \left(\frac{R_T^2}{R_L^2}\right)^{3/2} = \left(1 + \frac{2x}{R_T} + \frac{||\boldsymbol{\rho}||^2}{R_T^2}\right)^{-3/2} \approx 1 - \frac{3x}{R_T} \tag{2.35}$$

By substituting Eq. (2.35) into Eq. (2.34) and neglecting the second order small item, the linearized model can be derived as

$$\begin{aligned}\ddot{x} &= 2\ddot{u}_T\dot{y} + \dot{u}_T^2x + \ddot{u}_Ty + 2n_T^2x + a_x + a_R \\ \ddot{y} &= -2\ddot{u}_T\dot{x} + \dot{u}_T^2y - \ddot{u}_Tx - n_T^2y + a_y + a_S \\ \ddot{z} &= -n_T^2z + a_z + a_W\end{aligned} \tag{2.36}$$

Without the Lorentz acceleration, Eq. (2.36) is the well-known Tschanuer–Hempel (TH) equation [6], a dynamical model that describes the linearized relative orbital motion about an elliptic reference orbit.

Moreover, if the target is flying in a circular orbit, that is, $\dot{u}_T = \mu/R_T^3 = n_T$ and $\ddot{u}_T = 0$, then the linearized model further reduces to

$$\begin{aligned}\ddot{x} &= 2n_T\dot{y} + 3n_T^2x + a_x + a_R \\ \ddot{y} &= -2n_T\dot{x} + a_y + a_S \\ \ddot{z} &= -n_T^2z + a_z + a_W\end{aligned} \tag{2.37}$$

Also, without the Lorentz acceleration, Eq. (2.37) is the well-known Hill–Clohessy–Wiltshire (HCW) equation [7] that describes the linearized relative orbital motion about a circular reference orbit.

2.2.1.2 The Lorentz Force

A particle with a net charge of q_L is subject to the Lorentz force $\boldsymbol{F}_L$ when moving in a magnetic field $\boldsymbol{B}$ with a relative velocity of $\boldsymbol{V}_{\text{rel}}$, given by [8]

$$\boldsymbol{F}_L = q_L\boldsymbol{V}_{\text{rel}} \times \boldsymbol{B} \tag{2.38}$$

If the mass of the charged particle is m_L, the resulting Lorentz acceleration will be

$$\boldsymbol{a}_L = (q_L/m_L)\boldsymbol{V}_{\text{rel}} \times \boldsymbol{B} \tag{2.39}$$

To derive the expressions of the Lorentz acceleration in the LVLH frame, following assumptions are proposed: (1) the Earth magnetic field is modeled as a perfect dipole located at the Earth's center that is corotating with Earth; (2) the dipole is tilted by angle α with respect to the rotation axis of Earth, as shown in Fig. 2.8; (3) the Lorentz spacecraft could be regarded as a charged point mass [3].

As shown in Fig. 2.8, the transformation matrix from the ECI frame to the LVLH frame can be represented as

$$\boldsymbol{M}_{LI} = \boldsymbol{M}_3[u_T]\boldsymbol{M}_1[i_T]\boldsymbol{M}_3[\Omega_T] \tag{2.40}$$

where $\boldsymbol{M}_3[\cdot]$ and $\boldsymbol{M}_1[\cdot]$ are the basic transformation matrix, given by

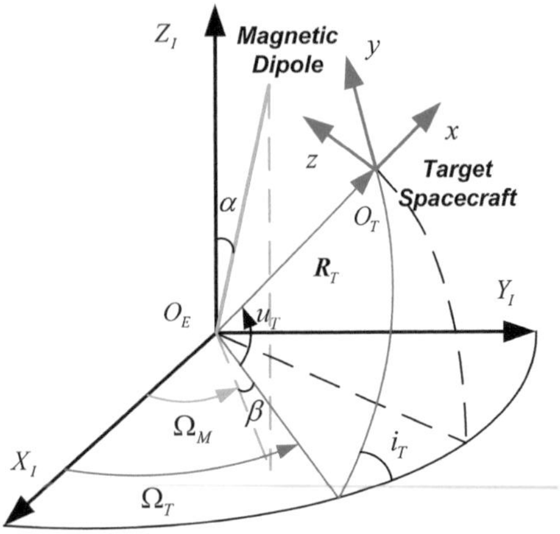

Fig. 2.8 Definitions of related angles. Reprinted from Ref. [10], Copyright 2014, with permission from SAGE

$$M_3[\cdot] = \begin{bmatrix} \cos(\cdot) & \sin(\cdot) & 0 \\ -\sin(\cdot) & \cos(\cdot) & 0 \\ 0 & 0 & 1 \end{bmatrix}, \quad M_1[\cdot] = \begin{bmatrix} 1 & 0 & 0 \\ 0 & \cos(\cdot) & \sin(\cdot) \\ 0 & -\sin(\cdot) & \cos(\cdot) \end{bmatrix} \tag{2.41}$$

Equation (2.12) has given the expression of $\hat{\boldsymbol{n}}$ in the ECI frame. Then, its expression in the LVLH frame can be obtained as $\hat{\boldsymbol{n}}^L = \boldsymbol{M}_{LI}\hat{\boldsymbol{n}}^I$, where the superscript L or I denotes the LVLH or ECI frame. By using Eqs. (2.12) and (2.40), the unit magnetic dipole vector can be expressed in the LVLH frame as

$$\begin{aligned} \hat{\boldsymbol{n}} &= \begin{bmatrix} n_x & n_y & n_z \end{bmatrix}^{\mathrm{T}} \\ &= \begin{bmatrix} -(\cos\beta\cos u_T + \sin\beta\cos i_T\sin u_T)\sin\alpha - \sin i_T\sin u_T\cos\alpha \\ (\cos\beta\sin u_T - \sin\beta\cos i_T\cos u_T)\sin\alpha - \sin i_T\cos u_T\cos\alpha \\ \sin\beta\sin i_T\sin\alpha - \cos i_T\cos\alpha \end{bmatrix} \end{aligned} \tag{2.42}$$

where the angle β is defined as $\beta = \Omega_M - \Omega_T$, as shown in Fig. 2.7.

Furthermore, the unit orbital radius vector of the Lorentz spacecraft can be expressed in the LVLH frame as

$$\hat{\boldsymbol{R}}_L = (1/R_L)\begin{bmatrix} R_T + x & y & z \end{bmatrix}^{\mathrm{T}} \tag{2.43}$$

Then, by substituting Eqs. (2.42) and (2.43) into Eq. (2.10), the magnetic field local at the Lorentz spacecraft is expressed in the LVLH frame as [9]

$$\boldsymbol{B} = \begin{bmatrix} B_x & B_y & B_z \end{bmatrix}^{\mathrm{T}} = (B_0/R_L^3)\left[3(\hat{\boldsymbol{n}}\cdot\hat{\boldsymbol{R}}_L)\hat{\boldsymbol{R}}_L - \hat{\boldsymbol{n}}\right] \tag{2.44}$$

Given that the geomagnetic field is corotating with Earth, the velocity of Lorentz spacecraft with respect to the local magnetic field is thus given by

$$\boldsymbol{V}_{\mathrm{rel}} = \frac{\mathrm{d}\boldsymbol{R}_L}{\mathrm{d}t} - \boldsymbol{\omega}_E \times \boldsymbol{R}_L = \dot{\boldsymbol{R}}_T + \dot{\boldsymbol{\rho}} + (\dot{\boldsymbol{u}}_T - \boldsymbol{\omega}_E) \times (\boldsymbol{R}_T + \boldsymbol{\rho}) \tag{2.45}$$

or in the LVLH coordinates

$$\boldsymbol{V}_{\mathrm{rel}} = \begin{bmatrix} V_x \\ V_y \\ V_z \end{bmatrix} = \begin{bmatrix} \dot{R}_T + \dot{x} - y(\dot{u}_T - \omega_E \cos i_T) - z\omega_E \sin i_T \cos u_T \\ \dot{y} + (R_T + x)(\dot{u}_T - \omega_E \cos i_T) + z\omega_E \sin i_T \sin u_T \\ \dot{z} + (R_T + x)\omega_E \sin i_T \cos u_T - y\omega_E \sin i_T \sin u_T \end{bmatrix} \tag{2.46}$$

Now, the expressions of the Lorentz acceleration in LVLH frame can be derived via substitution of Eqs. (2.44) and (2.46) into Eq. (2.39). Furthermore, the dynamical model of two-body Lorentz spacecraft relative motion can then be derived by substituting Eq. (2.39) into Eq. (2.34).

2.2.2 Analytical Solutions to Two-Body Model

2.2.2.1 Approximate Analytical Solutions

In this section, approximate analytical solutions are derived for a special case where the target is flying in an inclined circular Earth orbit. Since the reference orbit is circular, the linearized equations of Lorentz spacecraft relative motion have been given in Eq. (2.37).

To linearize the Lorentz acceleration and thereafter derive approximate analytical solutions to Eq. (2.37), following assumptions are proposed: (1) the relative coordinates between spacecraft are negligibly small as compared to the orbital radius of the target, that is, $|x|, |y|, |z| \ll R_T$; (2) the relative velocity between spacecraft are negligibly small as compared to the orbital translational velocity of the target, that is, $|\dot{x}|, |\dot{y}|, |\dot{z}| \ll \dot{u}_T R_T$ [9]; (3) except for the Lorentz acceleration, the control acceleration and other perturbations are neglected, that is, $\boldsymbol{a}_C = 0$. Based on these assumptions, the Lorentz acceleration can be linearized in the LVLH frame as [10]

$$a_x \approx (q_L/m_L)\left[R_T(n_T - \omega_E \cos i_T)B_z - (R_T\omega_E \sin i_T \cos u_T + \dot{z})B_y\right] \tag{2.47}$$

$$a_y \approx (q_L/m_L)(R_T\omega_E \sin i_T \cos u_T + \dot{z})B_x \tag{2.48}$$

$$a_z \approx c_1 \cos(u_T + \beta) + c_2 \cos(u_T - \beta) + c_3 \sin u_T \tag{2.49}$$

where the constant parameters c_1, c_2, and c_3 are given by

$$
\begin{aligned}
c_1 &= q_L B_0 (n_T - \omega_E \cos i_T) \sin \alpha (1 - \cos i_T)/(m_L R_T^2) \\
c_2 &= q_L B_0 (n_T - \omega_E \cos i_T) \sin \alpha (1 + \cos i_T)/(m_L R_T^2) \\
c_3 &= 2 q_L B_0 (n_T - \omega_E \cos i_T) \cos \alpha \, \sin i_T/(m_L R_T^2)
\end{aligned} \tag{2.50}
$$

As can be seen, the out-of-plane relative orbital motion is decoupled of the in-plane one. Thus, the equation of out-of-plane relative motion is solved at first. Denote z_0 and $\dot{z}_0$ as the initial relative normal position and velocity, respectively, and u_{T0} as the initial argument of latitude of the target, then the solution can be expressed as [10]

$$
\begin{bmatrix} z \\ \dot{z} \end{bmatrix} = \mathbf{\Phi}^{\text{out}} \begin{bmatrix} z_0 \\ \dot{z}_0 \end{bmatrix} + \begin{bmatrix} \lambda_z(t) \\ \dot{\lambda}_z(t) \end{bmatrix} \tag{2.51}
$$

with

$$
\mathbf{\Phi}^{\text{out}} = \begin{bmatrix} \cos n_T t & (\sin n_T t)/n_T \\ -n_T \sin n_T t & \cos n_T t \end{bmatrix} \tag{2.52}
$$

$$
\begin{aligned}
\lambda_z(t) = {} & k_{z1} \cos n_T t + \frac{1}{n_T} k_{z2} \sin n_T t - \frac{c_1 \cos[(n_T + \omega_E)t + \varphi_1]}{\omega_E(\omega_E + 2n_T)} \\
& - \frac{c_2 \cos[(n_T - \omega_E)t + \varphi_2]}{\omega_E(\omega_E - 2n_T)} - \frac{c_3 t}{2n_T} \cos(n_T t + \varphi_3) + \frac{c_3}{4n_T^2} \sin(n_T t + \varphi_3)
\end{aligned} \tag{2.53}
$$

where the coefficients k_{z1}, k_{z2}, φ_1, φ_2 and φ_3 are, respectively, given by

$$
\begin{aligned}
k_{z1} &= \frac{c_1 \cos \varphi_1}{\omega_E(\omega_E + 2n_T)} + \frac{c_2 \cos \varphi_2}{\omega_E(\omega_E - 2n_T)} - \frac{c_3}{4n_T^2} \sin \varphi_3 \\
k_{z2} &= -\frac{c_1(n_T + \omega_E)}{\omega_E(\omega_E + 2n_T)} \sin \varphi_1 - \frac{c_2(n_T - \omega_E)}{\omega_E(\omega_E - 2n_T)} \sin \varphi_2 + \frac{c_3}{4n_T} \cos \varphi_3
\end{aligned} \tag{2.54}
$$

$$
\begin{aligned}
\varphi_1 &= u_{T0} + \Omega_0 - \Omega_T \\
\varphi_2 &= u_{T0} - \Omega_0 + \Omega_T \\
\varphi_3 &= u_{T0}
\end{aligned} \tag{2.55}
$$

Substitution of the expression of $\dot{z}(t)$ into Eqs. (2.47) and (2.48) yields that [10]

$$
a_x = \kappa_{t,0}^x + \sum_{m=1}^{2} t^{m-1} \left[\sum_{i=1}^{9} C_{m,i}^x \sin(A_{m,i}^x t + \theta_{m,i}^x) + D_{m,i}^x \cos(B_{m,i}^x t + \phi_{m,i}^x) \right] \tag{2.56}
$$

$$a_y = \kappa_{t,0}^y + \kappa_{t,1}^y t + \sum_{m=1}^{2} t^{m-1} \left[\sum_{i=1}^{9} C_{m,i}^y \sin(A_{m,i}^y t + \theta_{m,i}^y) + D_{m,i}^y \cos(B_{m,i}^y t + \phi_{m,i}^y) \right] \tag{2.57}$$

where the related coefficients are summarized in the Appendix.

By substituting Eqs. (2.56) and (2.57) into Eq. (2.37), the approximate analytical expressions of the in-plane relative motion are derived as [10]

$$\begin{bmatrix} x \\ y \\ \dot{x} \\ \dot{y} \end{bmatrix} = \mathbf{\Phi}^{in} \begin{bmatrix} x_0 \\ y_0 \\ \dot{x}_0 \\ \dot{y}_0 \end{bmatrix} + \begin{bmatrix} \lambda_x(t) \\ \lambda_y(t) \\ \dot{\lambda}_x(t) \\ \dot{\lambda}_y(t) \end{bmatrix} \tag{2.58}$$

with

$$\mathbf{\Phi}^{in} = \begin{bmatrix} 4 - 3\cos n_T t & 0 & (\sin n_T t)/n_T & 2(1 - \cos n_T t) n_T \\ 6(\sin n_T t - n_T t) & 1 & 2(\cos n_T t - 1)/n_T & -3t + 4(\sin n_T t)/n_T \\ 3n_T \sin n_T t & 0 & \cos n_T t & 2\sin n_T t \\ 6n_T(\cos n_T t - 1) & 0 & -2\sin n_T t & 4\cos n_T t - 3 \end{bmatrix} \tag{2.59}$$

where x_0 and y_0 are initial relative radial and in-track position. $\dot{x}_0$ and $\dot{y}_0$ are initial relative radial and in-track velocity. The expressions of $\lambda_x(t)$, $\lambda_y(t)$, and related coefficients are summarized in the Appendix.

Heretofore, the approximate analytical expressions of the relative orbital motion of a Lorentz spacecraft about an inclined circular orbit have been derived as Eqs. (2.51) and (2.58). Without numerical integration, the relative states at an arbitrary time t can be determined analytically.

2.2.2.2 Numerical Simulations

In this section, a typical scenario in LEO is simulated to assess the accuracy of the analytical solutions. The initial orbit elements of the target spacecraft are listed in Table 2.1. The initial phase angle of the magnetic dipole is 40°.

In this scenario, the Lorentz spacecraft with a constant specific charge of 10^{-4} C/kg starts from the origin of the LVLH frame. The total simulation time is 15 orbital periods, that is, about 24 h. To evaluate the accuracy of the derived analytical solutions, the exact numerical solutions are introduced to make comparisons. Notably, the numerical solutions are derived via integration of the nonlinear equations of relative motion in Eq. (2.34). Also, to evaluate the effect of the dipole tilt angle on model accuracy, the analytical solutions for inclined orbit with non-tilted dipole proposed in [9] are also introduced to make comparisons. Define Δx, Δy, and Δz as the radial, in-track, and normal relative position error between the

Table 2.1 Initial orbit elements of the target (Reprinted from Ref. [10], Copyright 2014, with permission from SAGE)

Orbit element	Value
Semi-major axis (km)	6945.034
Eccentricity	0
Inclination (deg)	30
Right ascension of ascending node (deg)	30
Argument of latitude (deg)	20

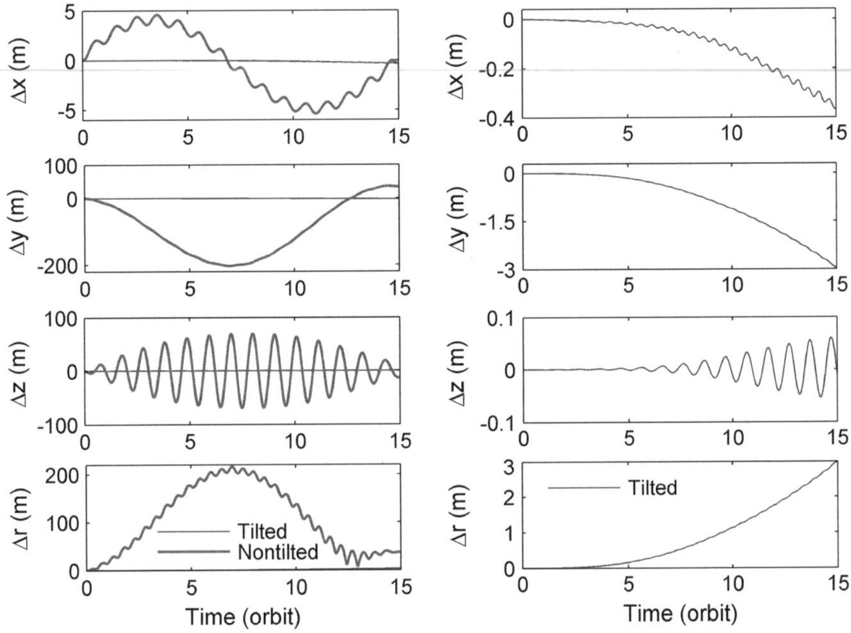

Fig. 2.9 Time histories of model errors for the tilted and nontilted dipole models. Reprinted from Ref. [10], Copyright 2014, with permission from SAGE

analytical and numerical solutions, respectively. Thus, the error relative distance is defined as $\Delta r = (\Delta x^2 + \Delta y^2 + \Delta z^2)^{1/2}$. Time histories of the model errors for the tilted and nontilted dipole models are then shown in Fig. 2.9. Also, details about the titled one are depicted separately in the right side of Fig. 2.9 to show it more distinctively. As can be seen, the tilted dipole model proposed in this book presents much more precise prediction of relative position than the nontilted one. For the tilted dipole model, the maximal relative position error is about 0.37, 2.97 and 0.06 m in each axis. However, for the nontilted dipole model, the model errors increase to 5.43, 203.95 and 69.69 m in each axis, which are about two orders of magnitude larger than those of the tilted model in y and z axes. This enhancement is due to the reason that the tilted dipole model explicitly includes the dipole tilt angle, which is more representative of the Earth's magnetic field. Thus, it presents enhanced accuracy. Meanwhile, it is notable that the model accuracy decreases with increasing simulation time and increasing specific charge. By using this numerical example, it has been proved that the dipole tile angle should be taken into account when analyzing the Lorentz-augmented relative motion about inclined LEOs.

2.2.3 J_2-Perturbed Model

2.2.3.1 Equations of J_2-Perturbed Relative Motion

J_2 perturbation, arising from the oblateness and nonhomogeneity of Earth, is one of the most dominant perturbations in LEOs. To capture the environment in LEOs more precisely, a nonlinear dynamical model of J_2-perturbed Lorentz spacecraft relative motion is developed in this subsection to analyze the effect of both Lorentz acceleration and J_2 perturbation on spacecraft relative motion. Notably, the reference orbit is a J_2-perturbed one. Therefore, a brief review of J_2-perturbed orbital dynamics of the target spacecraft is given at first. In the LVLH frame, the governing dynamics of the J_2-perturebd target's orbit can be descried by the following equations developed by Xu and Wang, given by [5]

$$\dot{R}_T = V_r \tag{2.60}$$

$$\dot{V}_r = -\frac{\mu}{R_T^2} + \frac{h_T^2}{R_T^3} - \frac{k_J}{R_T^4}(1 - 3\sin^2 i_T \sin^2 u_T) \tag{2.61}$$

$$\dot{h}_T = -\frac{k_J \sin^2 i_T \sin 2u_T}{R_T^3} \tag{2.62}$$

$$\dot{i}_T = -\frac{k_J \sin 2i_T \sin 2u_T}{2h_T R_T^3} \tag{2.63}$$

$$\dot{u}_T = \frac{h_T}{R_T^2} + \frac{2k_J \cos^2 i_T \sin^2 u_T}{h_T R_T^3} \tag{2.64}$$

$$\dot{\Omega}_T = -\frac{2k_J \cos i_T \sin^2 u_T}{h_T R_T^3} \tag{2.65}$$

where the six orbital variables $(R_T, V_r, h_T, i_T, u_T, \Omega_T)$ are, respectively, the orbital radius, radial velocity, orbital angular momentum, inclination, argument of latitude, and right ascension of the ascending node of the target spacecraft. The constant k_J is defined as

$$k_J = 3J_2\mu R_E^2/2 \tag{2.66}$$

In the presence of J_2 perturbations, the angular velocity of the LVLH frame $\boldsymbol{\omega}$ is given by [5]

$$\boldsymbol{\omega} = \begin{bmatrix} \omega_x \\ \omega_y \\ \omega_z \end{bmatrix} = \begin{bmatrix} -(k_J \sin 2i_T \sin u_T)/(h_T R_T^3) \\ 0 \\ h_T/R_T^2 \end{bmatrix} \tag{2.67}$$

Correspondingly, the angular acceleration of the LVLH frame can be derived as [5]

$$\boldsymbol{\varepsilon} = \dot{\boldsymbol{\omega}} = \begin{bmatrix} \varepsilon_x & 0 & \varepsilon_z \end{bmatrix}^{\mathrm{T}} \tag{2.68}$$

with

$$\begin{aligned} \varepsilon_x = & \frac{3V_r k_J \sin 2i_T \sin u_T}{R_T^4 h_T} - \frac{k_J \sin 2i_T \cos u_T}{R_T^5} \\ & - \frac{8k_J^2 \sin^3 i_T \cos i_T \sin^2 u_T \cos u_T}{R_T^6 h_T^2} \end{aligned} \tag{2.69}$$

$$\varepsilon_z = -\frac{2h_T V_r}{R_T^3} - \frac{k_J \sin^2 i_T \sin 2u_T}{R_T^5} \tag{2.70}$$

It is notable that both the angular velocity and acceleration are described in the LVLH frame.

Similarly, to derive the dynamical model of J_2-perturbed relative motion, following assumptions are proposed: (1) the Earth's magnetic field can be modeled as a tilted dipole located at the Earth's center corotating with Earth; (2) the dipole is tilted by angle α with respect to the rotation axis of Earth, as shown in Fig. 2.8; (3) the Lorentz spacecraft can be regarded as a charged point mass.

Based on these assumptions, the Lagrange dynamics method is introduced to develop the dynamical model in the LVLH frame. In this approach, the generalized coordinate vector $\boldsymbol{\chi}$ is chosen as the relative position between the Lorentz spacecraft and the target, given by $\boldsymbol{\chi} = \begin{bmatrix} x & y & z \end{bmatrix}^{\mathrm{T}}$. Then, it remains to represent the Lagrange function by the generalized coordinate and its first order time derivative.

Under the effect of J_2 perturbations, the kinetic energy (per unit mass) of the Lorentz spacecraft is

$$T_{J_2} = \frac{1}{2} \boldsymbol{V}_{J_2}^{\mathrm{T}} \boldsymbol{V}_{J_2} \tag{2.71}$$

where $\boldsymbol{V}_{J2}$ is the absolute velocity of the Lorentz spacecraft, given by

$$\boldsymbol{V}_{J_2} = \frac{\mathrm{d}\boldsymbol{R}_L}{\mathrm{d}t} = \dot{\boldsymbol{R}}_L + \boldsymbol{\omega} \times \boldsymbol{R}_L = \begin{bmatrix} \dot{R}_T + \dot{x} - \omega_z y \\ \dot{y} + \omega_z (R_T + x) - \omega_x z \\ \dot{z} + \omega_x y \end{bmatrix} \tag{2.72}$$

Also, the magnetic potential (per unit mass) of the Lorentz spacecraft is

$$U_{L,J_2} = -\lambda_L \boldsymbol{V}_{rel,J_2}^{\mathrm{T}} \boldsymbol{A} = -\frac{q_L}{m_L} \boldsymbol{V}_{rel,J_2}^{\mathrm{T}} \left[(B_0 / R_L^2)(\hat{\boldsymbol{n}} \times \hat{\boldsymbol{R}}_L) \right] \tag{2.73}$$

where the unit magnetic dipole momentum vector is given in Eq. (2.42), and the unit orbital radius vector is given in Eq. (2.43). In the presence of J_2 perturbation, the velocity of the Lorentz spacecraft with respect to the local magnetic field is revised as

$$\begin{aligned} \boldsymbol{V}_{rel,J_2} &= \boldsymbol{V}_{J_2} - \boldsymbol{\omega}_E \times \boldsymbol{R}_L = \begin{bmatrix} V_{x,J_2} & V_{y,J_2} & V_{z,J_2} \end{bmatrix}^{\mathrm{T}} \\ &= \begin{bmatrix} \dot{R}_T + \dot{x} - y(\omega_z - \omega_E \cos i_T) - z\omega_E \sin i_T \cos u_T \\ \dot{y} + (R_T + x)(\omega_z - \omega_E \cos i_T) - z(\omega_x - \omega_E \sin i_T \sin u_T) \\ \dot{z} + (R_T + x)\omega_E \sin i_T \cos u_T + y(\omega_x - \omega_E \sin i_T \sin u_T) \end{bmatrix} \end{aligned} \tag{2.74}$$

Furthermore, the gravitational potential (per unit mass) of the Lorentz spacecraft, including J_2 perturbation, is [5]

$$U_{g,J_2} = -\frac{\mu}{R_L} - \frac{k_J}{3R_L^3}(1 - 3\cos^2\theta) \tag{2.75}$$

with

$$\cos\theta = R_{Lz} / R_L \tag{2.76}$$

where R_{Lz} is the projection of $\boldsymbol{R}_L$ on the Z_I axis of the ECI frame, given by

$$R_{Lz} = (R_T + x)\sin i_T \sin u_T + y \sin i_T \cos u_T + z \cos i_T \tag{2.77}$$

Substitution of Eqs. (2.76) and (2.77) into Eq. (2.75) yields the gravitational potential in terms of the generalized coordinates, given by

$$U_{g,J_2} = -\frac{k_J}{3R_L^3}\left\{1 - 3\frac{\left[(R_T + x)\sin i_T \sin u_T + y\,\sin i_T \cos u_T + z\,\cos i_T\right]^2}{R_L^2}\right\} - \frac{\mu}{R_L} \tag{2.78}$$

Thus, the Lagrange function can be summarized as

$$L = T_{J_2} - U_{L,J_2} - U_{g,J_2} \tag{2.79}$$

where the expressions of T_{J_2}, U_{L,J_2}, and U_{g,J_2} are, respectively, shown in Eqs. (2.71), (2.73), and (2.78).

In consideration of the control acceleration $\boldsymbol{a}_C$, the Lagrangian equation for Lorentz spacecraft relative motion is

$$\frac{\mathrm{d}}{\mathrm{d}t}\left(\frac{\partial L}{\partial \dot{\boldsymbol{\chi}}}\right) - \frac{\partial L}{\partial \boldsymbol{\chi}} = \boldsymbol{a}_C \tag{2.80}$$

where $\dot{\boldsymbol{\chi}} = \begin{bmatrix}\dot{x} & \dot{y} & \dot{z}\end{bmatrix}^{\mathrm{T}}$. $\boldsymbol{a}_C = \begin{bmatrix}a_R & a_S & a_W\end{bmatrix}^{\mathrm{T}}$ is the control acceleration acting on the Lorentz spacecraft.

Solving the above Lagrangian function yields that [11]

$$\begin{aligned}\ddot{x} = {} & 2\dot{y}\omega_z - x(\eta_L^2 - \omega_z^2) + y\varepsilon_z - z\omega_x\omega_z - (\xi_L - \xi)\sin i_T \sin u_T \\ & - R_T(\eta_L^2 - \eta^2) + a_x + a_R\end{aligned} \tag{2.81}$$

$$\begin{aligned}\ddot{y} = {} & -2\dot{x}\omega_z + 2\dot{z}\omega_x - x\varepsilon_z - y(\eta_L^2 - \omega_z^2 - \omega_x^2) + z\varepsilon_x \\ & - (\xi_L - \xi)\sin i_T \cos u_T + a_y + a_S\end{aligned} \tag{2.82}$$

$$\ddot{z} = -2\dot{y}\omega_x - x\omega_x\omega_z - y\varepsilon_x - z(\eta_L^2 - \omega_x^2) - (\xi_L - \xi)\cos i_T + a_z + a_W \tag{2.83}$$

with

$$\eta_L^2 = \frac{\mu}{R_L^3} + \frac{k_J}{R_L^5} - \frac{5k_J R_{Lz}^2}{R_L^7} \tag{2.84}$$

$$\eta^2 = \frac{\mu}{R_T^3} + \frac{k_J}{R_T^5} - \frac{5k_J \sin^2 i_T \sin^2 u_T}{R_T^5} \tag{2.85}$$

$$\xi_L = \frac{2k_J R_{Lz}}{R_L^5} \tag{2.86}$$

$$\xi = \frac{2k_J \sin i_T \sin u_T}{R_T^4} \tag{2.87}$$

where $\boldsymbol{a}_L = \begin{bmatrix} a_x & a_y & a_z \end{bmatrix}^{\mathrm{T}}$ is the Lorentz acceleration acting on the Lorentz spacecraft, given by

$$\boldsymbol{a}_L = \lambda \boldsymbol{l} = \lambda \begin{bmatrix} l_x & l_y & l_z \end{bmatrix}^{\mathrm{T}} \tag{2.88}$$

with

$$\begin{aligned} l_x &= \frac{B_0}{R_L^5} V_{y,J_2} \{(3z^2 - R_L^2)n_z + 3z[(R_T + x)n_x + yn_y]\} \\ &\quad - \frac{B_0}{R_L^5} V_{z,J_2} \{(3y^2 - R_L^2)n_y + 3y[(R_T + x)n_x + zn_z]\} \end{aligned} \tag{2.89}$$

$$\begin{aligned} l_y &= \frac{B_0}{R_L^5} V_{z,J_2} \left\{[3(R_T + x)^2 - R_L^2]n_z + 3(R_T + x)(yn_y + zn_z)\right\} \\ &\quad - \frac{B_0}{R_L^5} V_{x,J_2} \{(3z^2 - R_L^2)n_z + 3z[(R_T + x)n_x + zn_z]\} \end{aligned} \tag{2.90}$$

$$\begin{aligned} l_z &= \frac{B_0}{R_L^5} V_{x,J_2} \{(3y^2 - R_L^2)n_y + 3y[(R_T + x)n_x + zn_z]\} \\ &\quad - \frac{B_0}{R_L^5} V_{y,J_2} \left\{[3(R_T + x)^2 - R_L^2]n_z + 3(R_T + x)(yn_y + zn_z)\right\} \end{aligned} \tag{2.91}$$

where $\hat{\boldsymbol{n}} = \begin{bmatrix} n_x & n_y & n_z \end{bmatrix}^{\mathrm{T}}$ and $\boldsymbol{V}_{\mathrm{rel},J_2} = \begin{bmatrix} V_{x,J_2} & V_{y,J_2} & V_{z,J_2} \end{bmatrix}^{\mathrm{T}}$ are, respectively, given in Eqs. (2.42) and (2.74). $(R_T, V_r, h_T, i_T, u_T, \Omega_T)$ are the solutions to Eqs. (2.60)–(2.65).

Heretofore, the dynamical model that describes the relative orbital motion of a Lorentz spacecraft about a J_2-perturbed reference orbit has been derived as Eqs. (2.81)–(2.91).

2.2.3.2 Numerical Simulations

A typical scenario in LEO is simulated to evaluate effect of J_2 perturbation on the Lorentz-augmented relative motion. The initial orbit elements of the target are chosen the same as those given in Table 2.1. Also, other simulation parameters are chosen the same as those given in Sect. 2.2.2.2. The only difference is the inclusion of J_2 perturbation here.

The exact trajectories of the relative position are generated by numerical integrations of the nonlinear equations of J_2-perturbed relative motion from Eqs. (2.81) to (2.83). For the tilted and nontilted dipole models, time histories of model errors

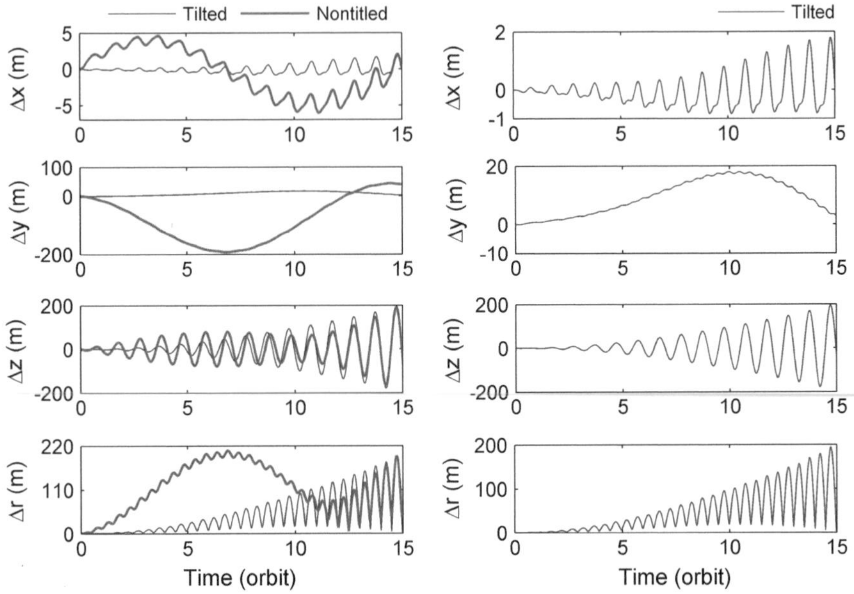

Fig. 2.10 Time histories of model errors for the titled and nontilted dipole modes in J_2-perturbed environment. Reprinted from Ref. [10], Copyright 2014, with permission from SAGE

are compared in Fig. 2.10. Also, details about the tilted dipole model are depicted separately in the right side of Fig. 2.10 to show it more distinctively. As can be seen, because both models do not take J_2 perturbation into account, the prediction accuracies of both models decrease distinctively after several periods in J_2-pertured environment, especially the relative position accuracy in z axis. Thus, it can be concluded that J_2 perturbation should be taken into account when analyzing the long-term Lorentz-augmented relative motion about LEOs.

2.3 Conclusions

Based on the assumption that the Earth's magnetic field can be approximated by a tilted dipole located at the Earth's center corotating with Earth, dynamical models of absolute and relative orbital motion of Lorentz spacecraft are developed by the Lagrange dynamics method in both two-body and J_2-perturbed environment. For the absolute orbital motion, the generalized coordinate is selected as the absolute position of the Lorentz spacecraft in space. For the relative one, the generalized coordinate is the relative position. All models in this chapter explicitly incorporate the dipole tilt angle, and approximate analytical solutions are derived for Lorentz-augmented spacecraft relative motion about circular inclined LEOs. Due to the inclusion of dipole tilt angle, which is more representative of the Earth's magnetic field, the newly derived analytical solutions present enhanced accuracy than previous nontilted one. Notably, numerical simulation results indicate that J_2 perturbation, one of the most dominant disturbances in LEOs, should be taken into account when analyzing the long-term Lorentz-augmented relative orbital motion.

References

1. Finlay CC, Maus S, Beggan CD et al (2010) International geomagnetic reference field: the eleventh generation. Geophys J Int 183:1216–1230
2. Wang X, Wang B, Li H (2012) An autonomous navigation scheme based on geomagnetic and starlight for small satellites. Acta Astronaut 81:40–50
3. Pollock GE, Gangestad JW, Longuski JM (2010) Inclination change in low-Earth orbit via the geomagnetic Lorentz force. J Guid Control Dyn 33:1387–1395
4. Greenwood DT (1997) Classical dynamics. Dover, New York
5. Xu G, Wang D (2008) Nonlinear dynamic equations of satellite relative motion around an oblate Earth. J Guid Control Dyn 31:1521–1524
6. Tschauner J, Hempel P (1965) Rendezvous with a target in an elliptical orbit. Acta Astronaut 11:104–109
7. Clohessy W, Wiltshire R (1960) Terminal guidance system for satellite rendezvous. J Aerosp Eng 27:653–658
8. Griffiths DJ (1999) Introduction to electrodynamics. Prentice-Hall, New Jersey
9. Pollock GE, Gangestad JW, Longuski JM (2011) Analytical solutions for the relative motion of spacecraft subject to Lorentz-force perturbations. Acta Astronaut 68:204–217
10. Huang X, Yan Y, Zhou Y et al (2014) Improved analytical solutions for relative motion of Lorentz spacecraft with application to relative navigation in low Earth orbit. Proc Inst Mech Eng Part G J Aerosp Eng 228:2138–2154. doi:10.1177/0954410013511426
11. Huang X, Yan Y, Zhou Y et al (2015) Nonlinear relative dynamics of Lorentz spacecraft about J_2-perturbed orbit. Proc Inst Mech Eng Part G J Aerosp Eng 229:467–478. doi:10.1177/0954410014537231

Chapter 3
Relative Navigation of Lorentz-Augmented Orbital Motion

The Lorentz acceleration acting on the Lorentz spacecraft is determined by the local magnetic field and the vehicle's velocity with respect to the local magnetic field. As derived in the last chapter, both the local magnetic field and the relative velocity are functions of the relative states between spacecraft, that is, relative position and velocity. Therefore, by incorporating the Lorentz acceleration into the dynamical model of relative orbital motion, the nonlinearity and coupling of system dynamics have been greatly increased. In view of this fact, it is necessary to design new relative navigation algorithms for Lorentz-augmented relative orbital motion. By using the dynamical model developed in the last chapter and the line of sight (LOS) and gyro measurement models, this chapter proposed both EKF and UKF algorithms specially for Lorentz spacecraft. Also, a typical scenario is simulated to compare the performance of these two filters.

3.1 State and Observation Equations

The relative motion between spacecraft consists of relative translational motion and relative rotational motion. To describe the translational and rotational motion of the Lorentz spacecraft with respect to the target spacecraft, the related coordinate frames are defined as follows [1].

As shown in Fig. 3.1, $O_E X_I Y_I Z_I$ is an Earth-centered inertial (ECI) frame. $O_T xyz$ is a local horizontal local vertical (LVLH) frame located at the center of mass (c.m.) of the target, where x axis is along the radial direction of the target, z axis is normal to the target's orbital plane, and y axis completes the right-handed Cartesian frame. $O_L X_B^L Y_B^L Z_B^L$ is the body frame of the Lorentz spacecraft, of which the axes are aligned with principle axes of inertial. O_L is the c.m. of the Lorentz spacecraft. Furthermore, it is assumed that the target's body frame is coincided with the LVLH frame.

© Springer Science+Business Media Singapore 2017

Y. Yan et al., *Dynamics and Control of Lorentz-Augmented Spacecraft Relative Motion*, DOI 10.1007/978-981-10-2603-4_3

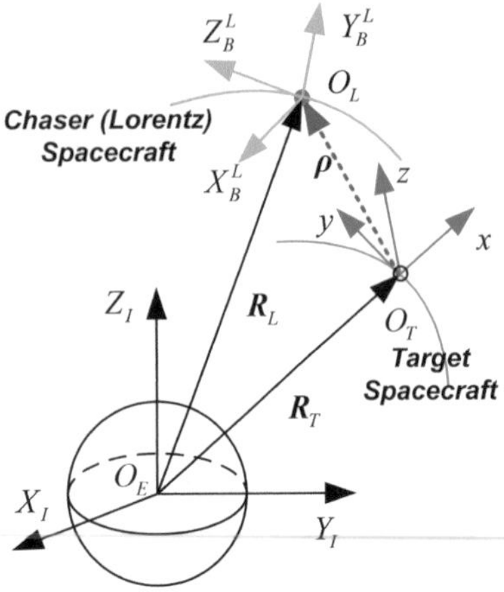

Fig. 3.1 Definitions of coordinate frames. Reprinted from Ref. [1], Copyright 2014, with permission from SAGE

The orbital radius vector of the Lorentz spacecraft and the target spacecraft are, respectively, $\boldsymbol{R}_L = \begin{bmatrix} R_T + x & y & z \end{bmatrix}^{\mathrm{T}}$ and $\boldsymbol{R}_T = \begin{bmatrix} R_T & 0 & 0 \end{bmatrix}^{\mathrm{T}}$. Then, the relative position vector between these two spacecraft is $\boldsymbol{\rho} = \boldsymbol{R}_L - \boldsymbol{R}_T = \begin{bmatrix} x & y & z \end{bmatrix}^{\mathrm{T}}$.

3.1.1 *State Equation of Relative Translational Motion*

The relative translational dynamics of Lorentz spacecraft has been detailedly discussed in Sect. 2.2. The equations of relative translational motion are given by

$$\begin{aligned} \ddot{x} &= 2\dot{u}_T\dot{y} + \dot{u}_T^2 x + \ddot{u}_T y + n_T^2 R_T - n_L^2(R_T + x) + a_x \\ \ddot{y} &= -2\dot{u}_T\dot{x} + \dot{u}_T^2 y - \ddot{u}_T x - n_L^2 y + a_y \\ \ddot{z} &= -n_L^2 z + a_z \end{aligned} \tag{3.1}$$

where $n_T = \sqrt{\mu/R_T^3}$ and $n_L = \sqrt{\mu/R_L^3}$. μ is the gravitational parameter of Earth. R_T and $R_L = \left[(R_T + x)^2 + y^2 + z^2\right]^{1/2}$ refer to the orbital radius of the target and the Lorentz spacecraft, respectively. u_T is the argument of latitude of the target. Then, $\dot{u}_T$ and $\ddot{u}_T$ refer to the orbital angular velocity and acceleration of the target, respectively. $\boldsymbol{a}_L = \begin{bmatrix} a_x & a_y & a_z \end{bmatrix}^{\mathrm{T}}$ is the Lorentz acceleration acting on the Lorentz spacecraft, given by

$$\boldsymbol{a}_L = (q_L/m_L)\boldsymbol{V}_{\mathrm{rel}} \times \boldsymbol{B} \tag{3.2}$$

with

$$\boldsymbol{V}_{\mathrm{rel}} = \begin{bmatrix} \dot{R}_T + \dot{x} - y(\dot{u}_T - \omega_E \cos i_T) - z\omega_E \sin i_T \cos u_T \\ \dot{y} + (R_T + x)(\dot{u}_T - \omega_E \cos i_T) + z\omega_E \sin i_T \sin u_T \\ \dot{z} + (R_T + x)\omega_E \sin i_T \cos u_T - y\omega_E \sin i_T \sin u_T \end{bmatrix} \tag{3.3}$$

$$\boldsymbol{B} = (B_0/R_L^3)\left[3(\hat{\boldsymbol{n}} \cdot \hat{\boldsymbol{R}}_L)\hat{\boldsymbol{R}}_L - \hat{\boldsymbol{n}}\right] \tag{3.4}$$

where q_L/m_L is the specific charge (i.e., charge-to-mass ratio) of Lorentz spacecraft. $\boldsymbol{V}_{\mathrm{rel}}$ is the velocity of the Lorentz spacecraft with respect to the local magnetic field $\boldsymbol{B}$. $\hat{\boldsymbol{R}}_L$ is the unit orbital radius vector of the Lorentz spacecraft, and $\hat{\boldsymbol{n}}$ is the unit magnetic dipole moment vector, given by

$$\hat{\boldsymbol{n}} = \begin{bmatrix} -(\cos\beta \cos u_T + \sin\beta \cos i_T \sin u_T)\sin\alpha - \sin i_T \sin u_T \cos\alpha \\ (\cos\beta \sin u_T - \sin\beta \cos i_T \cos u_T)\sin\alpha - \sin i_T \cos u_T \cos\alpha \\ \sin\beta \sin i_T \sin\alpha - \cos i_T \cos\alpha \end{bmatrix} \tag{3.5}$$

where i_T is the orbital inclination of the target. Also, the angle β is defined as $\beta = \Omega_M - \Omega_T$, where Ω_T is the right ascension of the ascending node of the target, and Ω_M is the inertial phase angle of the magnetic dipole, given by

$$\Omega_M = \omega_E t + \Omega_0 \tag{3.6}$$

where ω_E is the rotational rate of Earth and t is the time. Ω_0 represents the initial phase angle of the dipole.

Define the state vector of the Lorentz spacecraft with respect to the target as $\boldsymbol{X} = \begin{bmatrix} \boldsymbol{\rho}^{\mathrm{T}} & \boldsymbol{v}^{\mathrm{T}} \end{bmatrix}^{\mathrm{T}}$, where $\boldsymbol{v} = \begin{bmatrix} \dot{x} & \dot{y} & \dot{z} \end{bmatrix}^{\mathrm{T}}$ denotes the relative velocity. Provided that the target flies in a circular orbit (i.e., $\dot{u}_T = n_T$ and $\ddot{u}_T = 0$) and that the relative distance between spacecraft is negligibly small as compared to the orbital radius of either spacecraft, then the relative translation motion can be written in state-space model as

$$\dot{\boldsymbol{X}} = \boldsymbol{A}_q \boldsymbol{X} + \boldsymbol{B}_q \tag{3.7}$$

with

$$\boldsymbol{A}_q = \begin{bmatrix} \boldsymbol{0}_{3\times3} & \boldsymbol{I}_{3\times3} \\ \boldsymbol{A}_{v\rho} & \boldsymbol{A}_{vv} \end{bmatrix} \tag{3.8}$$

$$\boldsymbol{A}_{v\rho} = \begin{bmatrix} 3n_T^2 & 0 & 0 \\ 0 & 0 & 0 \\ 0 & 0 & -n_T^2 \end{bmatrix}, \quad \boldsymbol{A}_{vv} = \begin{bmatrix} 0 & 2n_T & 0 \\ -2n_T & 0 & 0 \\ 0 & 0 & 0 \end{bmatrix} \tag{3.9}$$

$$\boldsymbol{B}_q = \begin{bmatrix} \boldsymbol{0}_{1\times3} & a_x & a_y & a_z \end{bmatrix}^{\mathrm{T}} \tag{3.10}$$

where $\boldsymbol{0}_{M\times N}$ and $\boldsymbol{I}_{M\times N}$ denote an $M \times N$ zero and identity matrix, respectively.

3.1.2 State Equation of Relative Rotational Motion

Define the quaternion $\boldsymbol{q} = [\boldsymbol{q}_{13}^{\mathrm{T}} \quad q_4]^{\mathrm{T}} = [q_1 \quad q_2 \quad q_3 \quad q_4]^{\mathrm{T}}$ as the relative attitude quaternion between the Lorentz spacecraft and the target, where $\boldsymbol{q}_{13}$ is the vector part of the quaternion, and q_4 is the scalar part of the quaternion. Then, the dynamics of $\boldsymbol{q}$ is governed by [2]

$$\dot{\boldsymbol{q}} = \frac{1}{2}\boldsymbol{\Xi}(\boldsymbol{q})\boldsymbol{\omega}_r \tag{3.11}$$

with

$$\boldsymbol{\Xi}(\boldsymbol{q}) = \begin{bmatrix} q_4 & -q_3 & q_2 \\ q_3 & q_4 & -q_1 \\ -q_2 & q_1 & q_4 \\ -q_1 & -q_2 & -q_3 \end{bmatrix} \tag{3.12}$$

where $\boldsymbol{\omega}_r$ is the relative angular velocity between the Lorentz spacecraft and the target, given by

$$\boldsymbol{\omega}_r = \boldsymbol{\omega}_L - \boldsymbol{A}(\boldsymbol{q})\boldsymbol{\omega}_T \tag{3.13}$$

where $\boldsymbol{\omega}_L$ and $\boldsymbol{\omega}_T$ are, respectively, the angular velocity of the Lorentz spacecraft and the target. $\boldsymbol{A}(\boldsymbol{q})$ is the coordinate transformation matrix from the target's body frame (i.e., LVLH frame) to the Lorentz spacecraft's body frame, given by

$$\boldsymbol{A}(\boldsymbol{q}) = \begin{bmatrix} q_1^2 - q_2^2 - q_3^2 + q_4^2 & 2(q_1q_2 + q_3q_4) & 2(q_1q_3 - q_2q_4) \\ 2(q_1q_2 - q_3q_4) & -q_1^2 + q_2^2 - q_3^2 + q_4^2 & 2(q_2q_3 + q_1q_4) \\ 2(q_1q_3 + q_2q_4) & 2(q_2q_3 - q_1q_4) & -q_1^2 - q_2^2 + q_3^2 + q_4^2 \end{bmatrix} \tag{3.14}$$

Define the sampling time as $\Delta t = t_{k+1} - t_k$. By assuming that $\boldsymbol{\omega}_L$ and $\boldsymbol{\omega}_T$ remain constant during the sampling interval, the approximate solution to Eq. (3.11) can be represented as [3]

$$\boldsymbol{q}_{k+1} = \boldsymbol{\Theta}(\boldsymbol{\omega}_{L_k})\boldsymbol{\Gamma}(\boldsymbol{\omega}_{T_k})\boldsymbol{q}_k \tag{3.15}$$

with

$$\boldsymbol{\Theta}(\boldsymbol{\omega}_{L_k}) = \begin{bmatrix} \cos(\|\boldsymbol{\omega}_{L_k}\|\Delta t/2)\boldsymbol{I}_{3\times3} - [\boldsymbol{\psi}_k \times] & \boldsymbol{\psi}_k \\ -\boldsymbol{\psi}_k^{\mathrm{T}} & \cos(\|\boldsymbol{\omega}_{L_k}\|\Delta t/2) \end{bmatrix} \tag{3.16}$$

$$\boldsymbol{\Gamma}(\boldsymbol{\omega}_{T_k}) = \begin{bmatrix} \cos(||\boldsymbol{\omega}_{T_k}||\Delta t/2)\boldsymbol{I}_{3\times 3} - [\boldsymbol{\zeta}_k \times] & -\boldsymbol{\zeta}_k \\ \boldsymbol{\zeta}_k^{\mathrm{T}} & \cos(||\boldsymbol{\omega}_{T_k}||\Delta t/2) \end{bmatrix} \tag{3.17}$$

$$\boldsymbol{\psi}_k = \frac{\sin(||\boldsymbol{\omega}_{L_k}||\Delta t/2)\boldsymbol{\omega}_{L_k}}{||\boldsymbol{\omega}_{L_k}||}, \quad \boldsymbol{\zeta}_k = \frac{\sin(||\boldsymbol{\omega}_{T_k}||\Delta t/2)\boldsymbol{\omega}_{T_k}}{||\boldsymbol{\omega}_{L_k}||} \tag{3.18}$$

where $[\boldsymbol{a}\times]$ is a cross-product matrix defined as

$$[\boldsymbol{a}\times] = \begin{bmatrix} 0 & -a_3 & a_2 \\ a_3 & 0 & -a_1 \\ -a_2 & a_1 & 0 \end{bmatrix} \tag{3.19}$$

3.1.3 Observation Equation

A brief introduction of the line of sight (LOS) measurements and the gyro model is reviewed in this section.

As shown in Fig. 3.2, the beacons are fixed at known positions on the target, and the position sensing diode (PSD) is fixed on the Lorentz spacecraft. To derive the observation equation, following assumptions are imposed: (1) the sensor focal plane is known within the body frame of Lorentz spacecraft; (2) the target's body frame is coincided with the LVLH frame; (3) the z-axis of the sensor coordinate system is directed outward the boresight [3]. Then, if the ith beacon located at the known position (X_i, Y_i, Z_i) is observed in the focal plane as (χ_i, γ_i), the noiseless object to image space projection transformation can be represented as [4]

$$\chi_i = -f_d \frac{A_{11}(X_i - x) + A_{12}(Y_i - y) + A_{13}(Z_i - z)}{A_{31}(X_i - x) + A_{32}(Y_i - y) + A_{33}(Z_i - z)} \quad i = 1, 2, \ldots, N \tag{3.20}$$

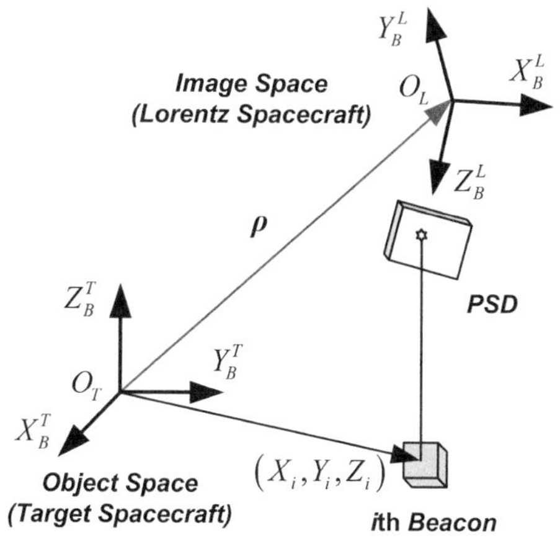

Fig. 3.2 Vision-based navigation system. Reprinted from Ref. [1], Copyright 2014, with permission from SAGE

$$\gamma_i = -f_d \frac{A_{21}(X_i - x) + A_{22}(Y_i - y) + A_{23}(Z_i - z)}{A_{31}(X_i - x) + A_{32}(Y_i - y) + A_{33}(Z_i - z)} \quad i = 1, 2, \ldots, N \tag{3.21}$$

where N is the number of the beacons, f_d is the known focal length, and A_{jk} is the unknown elements of the attitude matrix $\boldsymbol{A}(\boldsymbol{q})$.

Rewrite the ideal observation equation in a unit-vector form as [3]

$$\boldsymbol{b}_i = \boldsymbol{A}\boldsymbol{r}_i \quad i = 1, 2, \ldots, N \tag{3.22}$$

where the unit vectors $\boldsymbol{b}_i$ and $\boldsymbol{r}_i$ are, respectively, given by

$$\boldsymbol{b}_i = \frac{1}{\sqrt{f_d^2 + \chi_i^2 + \gamma_i^2}} \begin{bmatrix} -\chi_i \\ -\gamma_i \\ f_d \end{bmatrix} \tag{3.23}$$

$$\boldsymbol{r}_i = \frac{1}{\sqrt{(X_i - x)^2 + (Y_i - y)^2 + (Z_i - z)^2}} \begin{bmatrix} X_i - x \\ Y_i - y \\ Z_i - z \end{bmatrix} \tag{3.24}$$

Due to the existence of observation noises, the actual observation model is [3]

$$\tilde{\boldsymbol{b}}_i = \boldsymbol{A}\boldsymbol{r}_i + \boldsymbol{\eta}_i \tag{3.25}$$

where $\tilde{\boldsymbol{b}}_i$ is the ith observation vector, and $\boldsymbol{\eta}_i$ is the ith noise vector satisfying

$$E[\boldsymbol{\eta}_i] = \boldsymbol{0}_{3\times1}, \quad E[\boldsymbol{\eta}_i\boldsymbol{\eta}_i^{\mathrm{T}}] = \sigma_i^2 \boldsymbol{I}_{3\times3} \tag{3.26}$$

where $E[\,]$ denotes the expectation.

Generally, rate integrating gyros are used to measure the angular velocity of spacecraft, and the gyro model is given by [3]

$$\tilde{\boldsymbol{\omega}} = \boldsymbol{\omega} + \boldsymbol{\beta} + \boldsymbol{\eta}_v \tag{3.27}$$

$$\dot{\boldsymbol{\beta}} = \boldsymbol{\eta}_u \tag{3.28}$$

where $\tilde{\boldsymbol{\omega}}$ is the measured angular velocity, $\boldsymbol{\omega}$ is the actual angular velocity, and $\boldsymbol{\beta}$ is the gyro drift. $\boldsymbol{\eta}_v$ and $\boldsymbol{\eta}_v$ are independent zero-mean Gaussian white noises, satisfying $E[\boldsymbol{\eta}_v\boldsymbol{\eta}_v^{\mathrm{T}}] = \sigma_v^2\boldsymbol{I}_{3\times3}$ and $E[\boldsymbol{\eta}_u\boldsymbol{\eta}_u^{\mathrm{T}}] = \sigma_u^2\boldsymbol{I}_{3\times3}$.

The observation equations can now be summarized as Eqs. (3.25) and (3.27).

3.2 Filtering Algorithm for Relative Navigation

3.2.1 EKF Algorithm

Define the state vector as $\boldsymbol{s} = [\,\boldsymbol{\rho}^{\mathrm{T}} \quad \boldsymbol{v}^{\mathrm{T}} \quad \boldsymbol{q}^{\mathrm{T}} \quad \boldsymbol{\beta}_T^{\mathrm{T}} \quad \boldsymbol{\beta}_L^{\mathrm{T}}\,]^{\mathrm{T}}$, where $\boldsymbol{\beta}_T$ and $\boldsymbol{\beta}_L$ refer to the gyro drift of the target spacecraft and Lorentz spacecraft, respectively. Suppose that the target flies in a circular orbit. Then, the state equations are

$$\dot{\boldsymbol{s}} = \boldsymbol{f}(\boldsymbol{s}) + \boldsymbol{W} \tag{3.29}$$

with

$$\boldsymbol{f}(\boldsymbol{s}) = \left[\boldsymbol{f}_\rho^{\mathrm{T}} \quad \boldsymbol{f}_v^{\mathrm{T}} \quad \boldsymbol{f}_q^{\mathrm{T}} \quad \boldsymbol{f}_{\beta_T}^{\mathrm{T}} \quad \boldsymbol{f}_{\beta_L}^{\mathrm{T}}\right]^{\mathrm{T}} \tag{3.30}$$

$$\boldsymbol{W} = [\,\boldsymbol{0}_{1\times 3} \quad \boldsymbol{w}_a^{\mathrm{T}} \quad \boldsymbol{0}_{1\times 4} \quad \boldsymbol{\eta}_{u,T}^{\mathrm{T}} \quad \boldsymbol{\eta}_{u,L}^{\mathrm{T}}\,]^{\mathrm{T}} \tag{3.31}$$

$$\begin{aligned}
\boldsymbol{f}_\rho &= \boldsymbol{v} \\
\boldsymbol{f}_v &= \boldsymbol{A}_{v\rho}\boldsymbol{\rho} + \boldsymbol{A}_{vv}\boldsymbol{v} + (q_L/m_L)[\boldsymbol{v} + (\boldsymbol{n}_T - \omega_E) \times (\boldsymbol{R}_T + \boldsymbol{\rho})] \times \boldsymbol{B} \\
\boldsymbol{f}_q &= \boldsymbol{\Xi}(\boldsymbol{q})\boldsymbol{\omega}_r/2 \\
\boldsymbol{f}_{\beta_T} &= \boldsymbol{0}_{3\times 1} \\
\boldsymbol{f}_{\beta_L} &= \boldsymbol{0}_{3\times 1}
\end{aligned} \tag{3.32}$$

where $\boldsymbol{w}_a = [\,w_x \quad w_y \quad w_z\,]^{\mathrm{T}}$ is the disturbance acceleration, which is assumed to be a zero-mean Gaussian white noise, satisfying $\boldsymbol{w}_a \sim N(0, \boldsymbol{\sigma}_a^2)$.

The error state vector is defined as $\delta\boldsymbol{s} = [\,\delta\boldsymbol{\rho}^{\mathrm{T}} \quad \delta\boldsymbol{v}^{\mathrm{T}} \quad \delta\boldsymbol{\alpha}^{\mathrm{T}} \quad \delta\boldsymbol{\beta}_T^{\mathrm{T}} \quad \delta\boldsymbol{\beta}_L^{T}\,]^{\mathrm{T}}$, where $\delta\boldsymbol{\alpha}$ is a small angle error correction, given by [3]

$$\delta\boldsymbol{q} = \begin{bmatrix} \delta\boldsymbol{\alpha}/2 \\ 1 \end{bmatrix} \tag{3.33}$$

where $\delta\boldsymbol{q} = [\,\delta\boldsymbol{q}_{13}^{\mathrm{T}} \quad \delta q_4\,]^{\mathrm{T}}$ is error quaternion, with the vector part of the quaternion being $\delta\boldsymbol{q}_{13} = [\,\delta q_1 \quad \delta q_2 \quad \delta q_3\,]^{\mathrm{T}}$.

Then, the dynamical equations of the error states can be derived as

$$\delta\dot{\boldsymbol{s}} = \boldsymbol{F}\delta\boldsymbol{s} + \boldsymbol{G}\boldsymbol{w} \tag{3.34}$$

where

$$\boldsymbol{F} = \begin{bmatrix} \mathbf{0}_{3\times3} & \boldsymbol{I}_{3\times3} & \mathbf{0}_{3\times3} & \mathbf{0}_{3\times3} & \mathbf{0}_{3\times3} \\ \boldsymbol{F}_{v\rho} & \boldsymbol{F}_{vv} & \mathbf{0}_{3\times3} & \mathbf{0}_{3\times3} & \mathbf{0}_{3\times3} \\ \mathbf{0}_{3\times3} & \mathbf{0}_{3\times3} & -[\hat{\omega}_L\times] & \boldsymbol{A}(\hat{\boldsymbol{q}}) & -\boldsymbol{I}_{3\times3} \\ \mathbf{0}_{3\times3} & \mathbf{0}_{3\times3} & \mathbf{0}_{3\times3} & \mathbf{0}_{3\times3} & \mathbf{0}_{3\times3} \\ \mathbf{0}_{3\times3} & \mathbf{0}_{3\times3} & \mathbf{0}_{3\times3} & \mathbf{0}_{3\times3} & \mathbf{0}_{3\times3} \end{bmatrix} \tag{3.35}$$

$$\boldsymbol{F}_{v\rho} = \boldsymbol{A}_{v\rho} - (q_L/m_L)[\boldsymbol{B}\times]\{[\boldsymbol{n}_T\times] - [\omega_E\times]\} \tag{3.36}$$

$$\boldsymbol{F}_{vv} = \boldsymbol{A}_{vv} - (q_L/m_L)[\boldsymbol{B}\times] \tag{3.37}$$

$$\boldsymbol{G} = \begin{bmatrix} \mathbf{0}_{3\times3} & \mathbf{0}_{3\times3} & \mathbf{0}_{3\times3} & \mathbf{0}_{3\times3} & \mathbf{0}_{3\times3} \\ \boldsymbol{I}_{3\times3} & \mathbf{0}_{3\times3} & \mathbf{0}_{3\times3} & \mathbf{0}_{3\times3} & \mathbf{0}_{3\times3} \\ \mathbf{0}_{3\times3} & \boldsymbol{A}(\hat{\boldsymbol{q}}) & -\boldsymbol{I}_{3\times3} & \mathbf{0}_{3\times3} & \mathbf{0}_{3\times3} \\ \mathbf{0}_{3\times3} & \mathbf{0}_{3\times3} & \mathbf{0}_{3\times3} & \boldsymbol{I}_{3\times3} & \mathbf{0}_{3\times3} \\ \mathbf{0}_{3\times3} & \mathbf{0}_{3\times3} & \mathbf{0}_{3\times3} & \mathbf{0}_{3\times3} & \boldsymbol{I}_{3\times3} \end{bmatrix} \tag{3.38}$$

$$\boldsymbol{w} = \begin{bmatrix} \boldsymbol{w}_a^{\mathrm{T}} & \boldsymbol{\eta}_{v,T}^{\mathrm{T}} & \boldsymbol{\eta}_{v,L}^{\mathrm{T}} & \boldsymbol{\eta}_{u,T}^{\mathrm{T}} & \boldsymbol{\eta}_{u,L}^{\mathrm{T}} \end{bmatrix}^{\mathrm{T}} \tag{3.39}$$

The state transition matrix for the linear system Eq. (3.34) can thus be derived as $\boldsymbol{\Phi} = \mathrm{e}^{\boldsymbol{F}\cdot t}$. If the sampling time is sufficiently small, the state transition matrix can also be approximated by $\boldsymbol{\Phi} \approx \boldsymbol{I}_{15\times15} + \boldsymbol{F}\Delta t$.

The covariance matrix of the process noise $\boldsymbol{w}$ is

$$\boldsymbol{Q} = \begin{bmatrix} \sigma_a^2\boldsymbol{I}_{3\times3} & \mathbf{0}_{3\times3} & \mathbf{0}_{3\times3} & \mathbf{0}_{3\times3} & \mathbf{0}_{3\times3} \\ \mathbf{0}_{3\times3} & \sigma_{v,T}^2\boldsymbol{I}_{3\times3} & \mathbf{0}_{3\times3} & \mathbf{0}_{3\times3} & \mathbf{0}_{3\times3} \\ \mathbf{0}_{3\times3} & \mathbf{0}_{3\times3} & \sigma_{v,L}^2\boldsymbol{I}_{3\times3} & \mathbf{0}_{3\times3} & \mathbf{0}_{3\times3} \\ \mathbf{0}_{3\times3} & \mathbf{0}_{3\times3} & \mathbf{0}_{3\times3} & \sigma_{u,T}^2\boldsymbol{I}_{3\times3} & \mathbf{0}_{3\times3} \\ \mathbf{0}_{3\times3} & \mathbf{0}_{3\times3} & \mathbf{0}_{3\times3} & \mathbf{0}_{3\times3} & \sigma_{u,L}^2\boldsymbol{I}_{3\times3} \end{bmatrix} \tag{3.40}$$

Define the discrete covariance matrix used in the EKF as $\tilde{\boldsymbol{Q}}_k$, calculated via

$$\tilde{\boldsymbol{Q}}_k = \int_{t_{k-1}}^{t_k} \boldsymbol{\Phi}\boldsymbol{G}\boldsymbol{Q}\boldsymbol{G}^{\mathrm{T}}\boldsymbol{\Phi}^{\boldsymbol{T}}\mathrm{d}t \tag{3.41}$$

Also, when the sampling time is sufficiently small, the matrix $\tilde{\boldsymbol{Q}}_k$ can be approximated by

$$\tilde{\boldsymbol{Q}}_k = \Delta t\boldsymbol{G}\boldsymbol{Q}\boldsymbol{G}^{\mathrm{T}} \tag{3.42}$$

Define the observation vector as $\boldsymbol{y} = \begin{bmatrix} \tilde{\boldsymbol{b}}_1^{\mathrm{T}} & \tilde{\boldsymbol{b}}_2^{\mathrm{T}} & \ldots & \tilde{\boldsymbol{b}}_N^{\mathrm{T}} \end{bmatrix}^{\mathrm{T}}$, then the observation equation can be represented as

$$\boldsymbol{y} = \boldsymbol{h}(\boldsymbol{s}) + \boldsymbol{\eta} = \begin{bmatrix} \boldsymbol{A}(\boldsymbol{q})\boldsymbol{r}_1 \\ \boldsymbol{A}(\boldsymbol{q})\boldsymbol{r}_2 \\ \vdots \\ \boldsymbol{A}(\boldsymbol{q})\boldsymbol{r}_N \end{bmatrix} + \begin{bmatrix} \boldsymbol{\eta}_1 \\ \boldsymbol{\eta}_2 \\ \vdots \\ \boldsymbol{\eta}_N \end{bmatrix} \tag{3.43}$$

Then, the observation matrix $\boldsymbol{H}$ can be obtained as

$$\boldsymbol{H} = \frac{\partial \boldsymbol{h}}{\partial \boldsymbol{s}} = \begin{bmatrix} \boldsymbol{A}(\boldsymbol{q})\frac{\partial \boldsymbol{r}_1}{\partial \boldsymbol{\rho}} & \boldsymbol{0}_{3\times3} & [\boldsymbol{A}(\boldsymbol{q})\boldsymbol{r}_1\times] & \boldsymbol{0}_{3\times3} & \boldsymbol{0}_{3\times3} \\ \boldsymbol{A}(\boldsymbol{q})\frac{\partial \boldsymbol{r}_2}{\partial \boldsymbol{\rho}} & \boldsymbol{0}_{3\times3} & [\boldsymbol{A}(\boldsymbol{q})\boldsymbol{r}_2\times] & \boldsymbol{0}_{3\times3} & \boldsymbol{0}_{3\times3} \\ \vdots & \vdots & \vdots & \vdots & \vdots \\ \boldsymbol{A}(\boldsymbol{q})\frac{\partial \boldsymbol{r}_N}{\partial \boldsymbol{\rho}} & \boldsymbol{0}_{3\times3} & [\boldsymbol{A}(\boldsymbol{q})\boldsymbol{r}_N\times] & \boldsymbol{0}_{3\times3} & \boldsymbol{0}_{3\times3} \end{bmatrix} \tag{3.44}$$

with

$$\frac{\partial \boldsymbol{r}_i}{\partial \boldsymbol{\rho}} = \frac{1}{\upsilon_i}\begin{bmatrix} -(Y_i-y)^2-(Z_i-z)^2 & (X_i-x)(Y_i-y) & (X_i-x)(Z_i-z) \\ (X_i-x)(Y_i-y) & -(X_i-x)^2-(Z_i-z)^2 & (Y_i-y)(Z_i-z) \\ (X_i-x)(Z_i-z) & (Y_i-y)(Z_i-z) & -(X_i-x)^2-(Y_i-y)^2 \end{bmatrix} \tag{3.45}$$

$$\upsilon_i = [(X_i-x)^2+(Y_i-y)^2+(Z_i-z)^2]^{3/2} \tag{3.46}$$

where (X_i, Y_i, Z_i) is the location of the ith beacon in the target's body frame.

The covariance matrix of the observation noise $\boldsymbol{\eta}$ is

$$\boldsymbol{R} = \begin{bmatrix} \sigma_1^2\boldsymbol{I}_{3\times3} & \boldsymbol{0}_{3\times3} & \cdots & \boldsymbol{0}_{3\times3} \\ \boldsymbol{0}_{3\times3} & \sigma_2^2\boldsymbol{I}_{3\times3} & \cdots & \boldsymbol{0}_{3\times3} \\ \vdots & \vdots & \ddots & \vdots \\ \boldsymbol{0}_{3\times3} & \boldsymbol{0}_{3\times3} & \cdots & \sigma_N^2\boldsymbol{I}_{3\times3} \end{bmatrix} \tag{3.47}$$

The EKF algorithm is now summarized in Table 3.1.

Table 3.1 EKF for relative state estimation

Initialization	$\hat{\boldsymbol{s}}_0^+ = \hat{\boldsymbol{s}}(t_0)$, $\boldsymbol{P}_0^+ = \boldsymbol{P}(t_0)$, $\hat{\boldsymbol{q}}_0^+ = \boldsymbol{q}(t_0)$
Gain	$\boldsymbol{K}_k = \boldsymbol{P}_k^-\boldsymbol{H}_k^{\mathrm{T}}(\hat{\boldsymbol{q}}_k^-)[\boldsymbol{H}_k(\hat{\boldsymbol{q}}_k^-)\boldsymbol{P}_k^-\boldsymbol{H}_k^{\mathrm{T}}(\hat{\boldsymbol{q}}_k^-)+\boldsymbol{R}]^{-1}$
Update	$\boldsymbol{P}_k^+ = [\boldsymbol{I}-\boldsymbol{K}_k\boldsymbol{H}_k(\hat{\boldsymbol{q}}_k^-)]\boldsymbol{P}_k^-$, $\delta\hat{\boldsymbol{s}}_k^+ = \boldsymbol{K}_k[\boldsymbol{y}_k-\boldsymbol{h}_k(\hat{\boldsymbol{q}}_k^-)]$ $\delta\hat{\boldsymbol{s}}_k^+ = [\,\delta\hat{\boldsymbol{\rho}}_k^{+\mathrm{T}} \quad \delta\hat{\boldsymbol{v}}_k^{+\mathrm{T}} \quad \delta\hat{\boldsymbol{\alpha}}_k^{+\mathrm{T}} \quad \delta\hat{\boldsymbol{\beta}}_{T_k}^{+\mathrm{T}} \quad \delta\hat{\boldsymbol{\beta}}_{L_k}^{+\mathrm{T}}\,]^{\mathrm{T}}$ $\hat{\boldsymbol{\rho}}_k^+ = \hat{\boldsymbol{\rho}}_k^- + \delta\hat{\boldsymbol{\rho}}_k^+$, $\hat{\boldsymbol{v}}_k^+ = \hat{\boldsymbol{v}}_k^- + \delta\hat{\boldsymbol{v}}_k^+$, $\hat{\boldsymbol{q}}_k^+ = \hat{\boldsymbol{q}}_k^- + \boldsymbol{\Xi}(\hat{\boldsymbol{q}}_k^-)\delta\boldsymbol{\alpha}_k^+/2$ $\hat{\boldsymbol{\beta}}_{T_k}^+ = \hat{\boldsymbol{\beta}}_{T_k}^- + \delta\hat{\boldsymbol{\beta}}_{T_k}^+$, $\hat{\boldsymbol{\beta}}_{L_k}^+ = \hat{\boldsymbol{\beta}}_{L_k}^- + \delta\hat{\boldsymbol{\beta}}_{L_k}^+$
Propagation	$\hat{\boldsymbol{\rho}}_{k+1}^- = \hat{\boldsymbol{\rho}}_k^+ + \int_{t_k}^{t_{k+1}} \dot{\boldsymbol{\rho}}\mathrm{d}t$, $\hat{\boldsymbol{v}}_{k+1}^- = \hat{\boldsymbol{v}}_k^+ + \int_{t_k}^{t_{k+1}} \dot{\boldsymbol{v}}\mathrm{d}t$ $\hat{\omega}_{T_k}^+ = \tilde{\omega}_{T_k} - \hat{\boldsymbol{\beta}}_{T_k}^+$, $\hat{\omega}_{L_k}^+ = \tilde{\omega}_{L_k} - \hat{\boldsymbol{\beta}}_{L_k}^+$, $\hat{\boldsymbol{q}}_{k+1}^- = \boldsymbol{\Theta}(\hat{\omega}_{L_k}^+)\boldsymbol{\Gamma}(\hat{\omega}_{T_k}^+)\hat{\boldsymbol{q}}_k^+$ $\boldsymbol{P}_{k+1}^- = \boldsymbol{\Phi}_k\boldsymbol{P}_k^+\boldsymbol{\Phi}_k^{\mathrm{T}} + \boldsymbol{Q}_k$

3.2.2 UKF Algorithm

To conduct the UKF algorithm, the state equation (i.e., Eq. (3.29)) and the observation equation (i.e., Eq. (3.43)) are first discretized by the 4th order Runge–Kutta method. The discrete state and observation equations are

$$\boldsymbol{s}_{k+1} = \boldsymbol{\varphi}(\boldsymbol{s}_k) + \boldsymbol{W}_k \tag{3.48}$$

$$\boldsymbol{y}_{k+1} = \boldsymbol{h}(\boldsymbol{s}_k) + \boldsymbol{\eta}_k \tag{3.49}$$

where $\boldsymbol{W}_k$ is the process noise with covariance $\boldsymbol{Q}_k$, and $\boldsymbol{\eta}_k$ is the measurement noise with covariance $\boldsymbol{R}_k$. The function $\boldsymbol{\varphi}(\boldsymbol{s}_k)$ is

$$\boldsymbol{\varphi}(\boldsymbol{s}_k) = \boldsymbol{s}_k + (\boldsymbol{K}_1 + 2\boldsymbol{K}_2 + 2\boldsymbol{K}_3 + \boldsymbol{K}_4)/6 \tag{3.50}$$

where

$$\begin{aligned} \boldsymbol{K}_1 &= \boldsymbol{f}(\boldsymbol{s}_k, t_k) \\ \boldsymbol{K}_2 &= \boldsymbol{f}(\boldsymbol{s}_k + \boldsymbol{K}_1 \cdot \Delta t/2, t_k + \Delta t/2) \\ \boldsymbol{K}_3 &= \boldsymbol{f}(\boldsymbol{s}_k + \boldsymbol{K}_2 \cdot \Delta t/2, t_k + \Delta t/2) \\ \boldsymbol{K}_4 &= \boldsymbol{f}(\boldsymbol{s}_k + \boldsymbol{K}_3 \cdot \Delta t, t_k + \Delta t) \end{aligned} \tag{3.51}$$

Note that the unscented transformation cannot guarantee the unit norm of the quaternion. Then, following the treatments in [5], a local error-quaternion $\delta\boldsymbol{p}$ is introduced, given by

$$\delta\boldsymbol{p} = f[\delta\boldsymbol{q}_{13}/(a + \delta q_4)] \tag{3.52}$$

where $0 < a < 1$ is a design constant, and f is a scalar factor. Here, f is chosen as $f = 2(a+1)$.

Now, the UKF algorithm is summarized in Table 3.2, where the state vector is $\boldsymbol{s}_k = [\boldsymbol{\rho}_k^{\mathrm{T}} \quad \boldsymbol{v}_k^{\mathrm{T}} \quad \delta\boldsymbol{p}_k^{\mathrm{T}} \quad \boldsymbol{\beta}_{T_k}^{\mathrm{T}} \quad \boldsymbol{\beta}_{L_k}^{\mathrm{T}}]^{\mathrm{T}}$, and $\boldsymbol{s}_k^a = [\boldsymbol{s}_k^{\mathrm{T}} \quad \boldsymbol{W}_k^{\mathrm{T}} \quad \boldsymbol{\eta}_k^{\mathrm{T}}]^{\mathrm{T}}$ is the augmented state vector. n is the dimension of the augmented state vector. κ is a scaling parameter, which is generally set as 0 or $n - 3$ [6]. ξ determines the distribution of the sigma points around $\hat{\boldsymbol{s}}_k$, which is generally set as a small positive constant that satisfies $\xi \in [10^{-4}, 1]$ [6]. ζ is dependent on the distribution. For Gaussian distributions, $\zeta = 2$ is optimal [6].

3.2.3 Numerical Simulations

A typical scenario of Lorentz-augmented spacecraft rendezvous is simulated here to verify and compare the performance of the EKF and UKF algorithms. The initial

Table 3.2 UKF for relative state estimation

Selection of parameters	$\kappa,\ \xi,\ \zeta$
Calculation of weight	$\lambda = \xi^2(n+\kappa) - n$ $W_0^M = \lambda/(n+\lambda),\ W_0^C = \lambda/(n+\lambda) + (1-\xi^2+\zeta)$ $W_i^M = W_i^C = 1/[2(n+\lambda)],\quad i = 1,2,\ldots,2n$
Initialization	$\hat{\mathbf{s}}_0 = E[\mathbf{s}_0],\ \boldsymbol{P}_0 = E[(\mathbf{s}_0 - \hat{\mathbf{s}}_0)(\mathbf{s}_0 - \hat{\mathbf{s}}_0)^{\mathrm{T}}]$ $\hat{\mathbf{s}}_0^a = [\hat{\mathbf{s}}_0^{\mathrm{T}}\quad \boldsymbol{0}^{\mathrm{T}}\quad \boldsymbol{0}^{\mathrm{T}}]^{\mathrm{T}},\ \boldsymbol{P}_0^a = \mathrm{diag}(\boldsymbol{P}_0, \boldsymbol{Q}, \boldsymbol{R})$ $\delta\hat{\boldsymbol{p}}_0 = \boldsymbol{0},\ \hat{\boldsymbol{\omega}}_{r0} = \hat{\boldsymbol{\omega}}_{L0} - \boldsymbol{A}(\hat{\boldsymbol{q}}_0)\hat{\boldsymbol{\omega}}_{T0}$
Calculation of sigma points	$\boldsymbol{\chi}_k^0 = \hat{\mathbf{s}}_k^a$ $\boldsymbol{\chi}_k^i = \hat{\mathbf{s}}_k^a + \sqrt{n+\lambda}\sqrt{\boldsymbol{P}_k^a},\quad i = 1,2,\ldots,n$ $\boldsymbol{\chi}_k^i = \hat{\mathbf{s}}_k^a - \sqrt{n+\lambda}\sqrt{\boldsymbol{P}_k^a},\quad i = n+1, n+2,\ldots,2n$ $\delta\boldsymbol{q}_{4_k}^{+} = (-a\Vert\delta\hat{\boldsymbol{p}}_{k+1}^{+}\Vert^2 + f\sqrt{f^2 + (1-a^2)\Vert\delta\hat{\boldsymbol{p}}_{k+1}^{+}\Vert^2})/(f^2 + \Vert\delta\hat{\boldsymbol{p}}_{k+1}^{+}\Vert^2),$ $i = 1,2,\ldots,2n$ $\delta\boldsymbol{q}_{13_k}^{i+} = f^{-1}(a + \delta q_{4_k}^{i+})\delta\hat{\boldsymbol{p}}_k^{i+},\quad i = 1,2,\ldots,2n$ $\hat{\boldsymbol{q}}_k^{0+} = \hat{\boldsymbol{q}}_k^{+},\ \hat{\boldsymbol{q}}_k^{i+} = \delta\boldsymbol{q}_k^{i+} \otimes \hat{\boldsymbol{q}}_k^{+},\quad i = 1,2,\ldots,2n$ $\hat{\boldsymbol{\omega}}_{T_k}^{i+} = \tilde{\boldsymbol{\omega}}_{T_k} - \hat{\boldsymbol{\beta}}_{T_k}^{i+},\ \hat{\boldsymbol{\omega}}_{L_k}^{i+} = \tilde{\boldsymbol{\omega}}_{L_k} - \hat{\boldsymbol{\beta}}_{L_k}^{i+}$
Time update equations	$\boldsymbol{\chi}_{k+1}^{i-} = \boldsymbol{\varphi}(\boldsymbol{\chi}_k^i, k)$ (except for $\boldsymbol{\chi}_{\delta\mathbf{p}_k}^i$) $\hat{\boldsymbol{q}}_{k+1}^{i-} = \boldsymbol{\Theta}(\hat{\boldsymbol{\omega}}_{L_k}^{i+})\boldsymbol{\Gamma}(\hat{\boldsymbol{\omega}}_{T_k}^{i+})\hat{\boldsymbol{q}}_k^{i+},\quad i = 0,1,2,\ldots,2n$ $\delta\boldsymbol{q}_{k+1}^{i-} = \hat{\boldsymbol{q}}_{k+1}^{i-} \otimes (\hat{\boldsymbol{q}}_{k+1}^{0-})^{-1},\quad i = 0,1,2,\ldots,2n$ $\boldsymbol{\chi}_{\delta\mathbf{p}_{k+1}}^{0-} = \boldsymbol{0},\ \boldsymbol{\chi}_{\delta\mathbf{p}_{k+1}}^{i-} = f\delta\boldsymbol{q}_{13_{k+1}}^{i-}/(a + \delta q_{4_{k+1}}^{i-}),\quad i = 1,2,\ldots,2n$ $\hat{\mathbf{s}}_{k+1}^{-} = \sum_{i=0}^{2n} W_i^M \boldsymbol{\chi}_{k+1}^{i-},$ $\boldsymbol{P}_{k+1}^{-} = \sum_{i=0}^{2n} W_i^C[(\boldsymbol{\chi}_{k+1}^{i-} - \hat{\mathbf{s}}_{k+1}^{-})(\boldsymbol{\chi}_{k+1}^{i-} - \hat{\mathbf{s}}_{k+1}^{-})^{\mathrm{T}}] + \boldsymbol{Q}_{k+1}$ $\boldsymbol{y}_{k+1}^{i-} = \boldsymbol{h}(\boldsymbol{\chi}_{k+1}^{i-}, k),\ \hat{\boldsymbol{y}}_{k+1}^{-} = \sum_{i=0}^{2n} W_i^M \boldsymbol{y}_{k+1}^{i-}$
Measurement update equations	$\boldsymbol{P}_{\hat{y}\hat{y}}^{k+1} = \sum_{i=0}^{2n} W_i^C(\boldsymbol{y}_{k+1}^{i-} - \hat{\boldsymbol{y}}_{k+1}^{-})(\boldsymbol{y}_{k+1}^{i-} - \hat{\boldsymbol{y}}_{k+1}^{-})^{\mathrm{T}} + \boldsymbol{R}_{k+1}$ $\boldsymbol{P}_{\hat{s}\hat{y}}^{k+1} = \sum_{i=0}^{2n} W_i^C(\boldsymbol{\chi}_{k+1}^{i-} - \hat{\mathbf{s}}_{k+1}^{-})(\boldsymbol{y}_{k+1}^{i-} - \hat{\boldsymbol{y}}_{k+1}^{-})^{\mathrm{T}},$ $\boldsymbol{K}_{k+1} = \boldsymbol{P}_{\hat{s}\hat{y}}^{k+1}(\boldsymbol{P}_{\hat{y}\hat{y}}^{k+1})^{-1}$ $\hat{\mathbf{s}}_{k+1}^{+} = \hat{\mathbf{s}}_{k+1}^{-} + \boldsymbol{K}_{k+1}(\boldsymbol{y}_{k+1} - \hat{\boldsymbol{y}}_{k+1}^{-}),\ \boldsymbol{P}_{k+1}^{+} = \boldsymbol{P}_{k+1}^{-} - \boldsymbol{K}_{k+1}\boldsymbol{P}_{\hat{y}\hat{y}}^{k+1}\boldsymbol{K}_{k+1}^{\mathrm{T}}$ $\delta\boldsymbol{q}_{4_{k+1}}^{+} = (-a\Vert\delta\hat{\boldsymbol{p}}_{k+1}^{+}\Vert^2 + f\sqrt{f^2 + (1-a^2)\Vert\delta\hat{\boldsymbol{p}}_{k+1}^{+}\Vert^2})/(f^2 + \Vert\delta\hat{\boldsymbol{p}}_{k+1}^{+}\Vert^2),$ $i = 1,2,\ldots,2n$ $\delta\boldsymbol{q}_{13_{k+1}}^{+} = f^{-1}(a + \delta q_{4_{k+1}}^{+})\delta\hat{\boldsymbol{p}}_{k+1}^{+},\quad i = 1,2,\ldots,2n$ $\hat{\boldsymbol{q}}_{k+1}^{+} = \delta\boldsymbol{q}_{k+1}^{+} \otimes \hat{\boldsymbol{q}}_{k+1}^{0-},\ \delta\hat{\boldsymbol{p}}_{k+1}^{+} = \boldsymbol{0}$ $\hat{\boldsymbol{\omega}}_{L_{k+1}} = \tilde{\boldsymbol{\omega}}_{L_{k+1}} - \hat{\boldsymbol{\beta}}_{L_{k+1}}^{+},\ \hat{\boldsymbol{\omega}}_{T_{k+1}} = \tilde{\boldsymbol{\omega}}_{T_{k+1}} - \hat{\boldsymbol{\beta}}_{T_{k+1}}^{+},$ $\hat{\boldsymbol{\omega}}_{r_{k+1}} = \hat{\boldsymbol{\omega}}_{L_{k+1}} - \boldsymbol{A}(\hat{\boldsymbol{q}}_{k+1}^{+})\hat{\boldsymbol{\omega}}_{T_{k+1}}$

Reprinted from Ref. [1], Copyright 2014, with permission from SAGE

orbital elements of the target spacecraft are listed in Table 3.3. The initial relative position and velocity vector between the Lorentz spacecraft and the target are $\boldsymbol{\rho}(0) = [-3.10\quad 266.57\quad -74.35]^{\mathrm{T}}$ m and $\boldsymbol{v}(0) = [0.10\quad 5.67\quad 62.36]^{\mathrm{T}} \times 10^{-3}$ m/s, respectively. By using the analytical solutions in Sect. 2.2.2.1, it can be derived that a propellant rendezvous at be achieved at the end of one orbital period if the Lorentz spacecraft maintains a constant specific charge of -1.8×10^{-4} C/kg.

The initial relative attitude quaternion is set as $\boldsymbol{q}(0) = \begin{bmatrix} 0 & 0 & 0 & 1 \end{bmatrix}^{\mathrm{T}}$, and the angular velocity of the Lorentz spacecraft and the target spacecraft are, respectively, set as $\boldsymbol{\omega}_L = \begin{bmatrix} -0.002 & 0 & 0.0011 \end{bmatrix}^{\mathrm{T}}$ rad/s and $\boldsymbol{\omega}_T = \begin{bmatrix} 0 & -0.001 & 0.0011 \end{bmatrix}^{\mathrm{T}}$ rad/s. The total simulation time is set as one orbital period, and the sampling time for all sensors are chosen as 0.5 s. The six beacons are fixed on the target spacecraft at the following positions:

$$\begin{array}{lll} X_1 = 0.5\,\mathrm{m}, & Y_1 = 0.5\,\mathrm{m}, & Z_1 = 0.0\,\mathrm{m} \\ X_2 = -0.5\,\mathrm{m}, & Y_2 = 0.5\,\mathrm{m}, & Z_2 = 0.0\,\mathrm{m} \\ X_3 = -0.5\,\mathrm{m}, & Y_3 = -0.5\,\mathrm{m}, & Z_3 = 0.0\,\mathrm{m} \\ X_4 = 0.5\,\mathrm{m}, & Y_4 = -0.5\,\mathrm{m}, & Z_4 = 0.0\,\mathrm{m} \\ X_5 = 0.3\,\mathrm{m}, & Y_5 = 0.6\,\mathrm{m}, & Z_5 = 0.1\,\mathrm{m} \\ X_6 = 0.0\,\mathrm{m}, & Y_6 = 0.3\,\mathrm{m}, & Z_6 = -0.2\,\mathrm{m} \end{array} \tag{3.53}$$

The rendezvous trajectory is shown in Fig. 3.3, and the remaining simulation parameters are summarized in Table 3.4.

The initial values of the states are set as the sum of the true values and the sampled noise values. Notably, the noise values are generated according to the corresponding variances listed in Table 3.4. Also, the initial covariance matrices are generated based on the corresponding variances listed in Table 3.4. For example, the initial covariance matrix of the relative position error is set as $5^2\boldsymbol{I}_{3\times3}$ m^2, and that of the relative attitude is set as $0.2^2\boldsymbol{I}_{3\times3}$ deg^2.

The simulation results are then shown from Figs. 3.4, 3.5, 3.6, 3.7 and 3.8. Notably, the state estimation errors and corresponding 3σ bounds are the mean results of 100 Monte Carol runs. Time histories of the relative position estimation errors are shown in Fig. 3.4, from which it is clear that the estimation errors of UKF are obviously smaller than those of EKF. It takes EKF about 0.7 T (orbital period) to converge to small estimation errors, but the convergent time for UKF is only about 0.3 T.

Furthermore, the estimation errors of UKF are always kept with the 3σ bounds, whereas those of EKF are not. Time histories of the relative velocity estimation errors are shown in Fig. 3.5. Similarly, UKF presents faster and more precise estimation than EKF. The simulation results indicate that due to the inclusion of nonlinear Lorentz acceleration, the nonlinearity of the relative dynamical model increases. Then, the estimation accuracy of EKF, which is more precise for a linear

Table 3.3 Initial orbit elements of the target

Orbit element	Value
Semi-major axis (km)	6945.034
Eccentricity	0
Inclination (deg)	30
Right ascension of ascending node (deg)	30
Argument of latitude (deg)	20

Reprinted from Ref. [1], Copyright 2014, with permission from SAGE

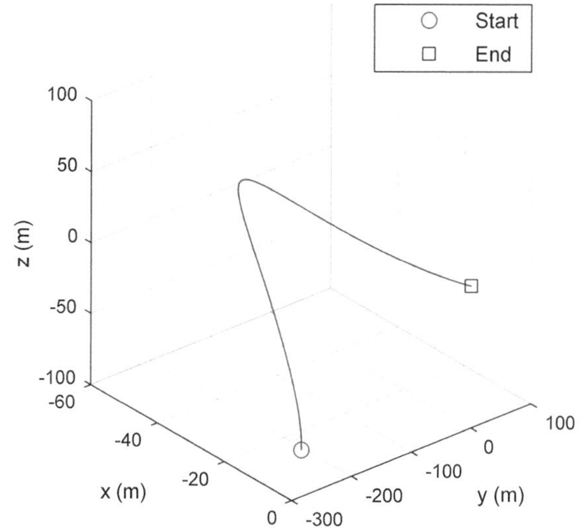

Fig. 3.3 Relative transfer trajectory Reprinted from Ref. [1], Copyright 2014, with permission from SAGE

Table 3.4 Simulation parameters

Parameter	Value
Initial error	
Relative position	$\sigma_\rho^r = 5$ m/axis
Relative velocity	$\sigma_v^r = 0.5$ (m/s)/axis
Relative attitude	$\sigma_a^r = 0.2$ deg/axis
Gyro drift	$\sigma_{\beta_L} = \sigma_{\beta_T} = 1/3$ deg/h/axis
Sensor error	
Process noise	$\sigma_a = \sqrt{10} \times 10^{-11}$ m/(s$^{3/2}$)
Gyro noise	$\sigma_{v_L} = \sigma_{v_T} = \sqrt{10} \times 10^{-7}$ rad/(s$^{1/2}$)
	$\sigma_{u_L} = \sigma_{u_T} = \sqrt{10} \times 10^{-10}$ rad/(s$^{3/2}$)
Gyro bias	$\boldsymbol{\beta}_L = \boldsymbol{\beta}_T = [1 \quad 1 \quad 1]^{\mathrm{T}}$ deg/h
VISNAV noise	$\sigma_i = 5 \times 10^{-4}$ deg

Reprinted from Ref. [1], Copyright 2014, with permission from SAGE

system, decreases. Figure 3.6 compares the estimation errors of relative attitude. As can be seen, the estimation accuracy of UKF is superior to that of EKF as a result of the inherent nonlinearity of relative attitude dynamics. Also, the attitude knowledge of UKF is within 0.2° on each axis. Estimates of the gyro drifts of the target spacecraft and Lorentz spacecraft are compared in Figs. 3.7 and 3.8, respectively. Likewise, despite that both filters could converge to the true values of gyro drifts, UKF presents faster convergent rate with enhanced precision. Aforementioned analyses verify the advantages of UKF over EKF in capturing the inherent nonlinearity of Lorentz-augmented relative orbital motion.

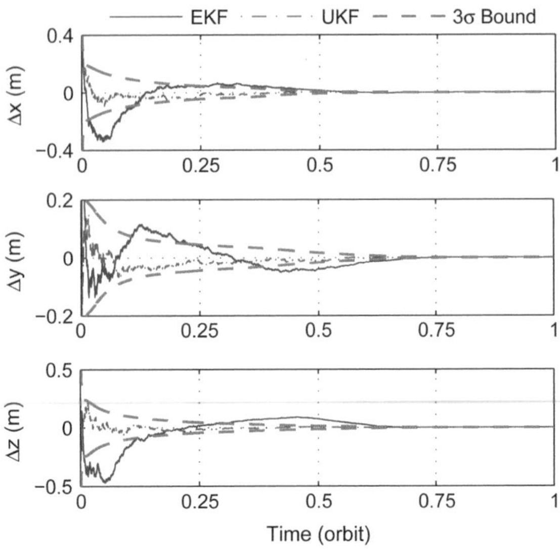

Fig. 3.4 Relative position estimation errors. Reprinted from Ref. [1], Copyright 2014, with permission from SAGE

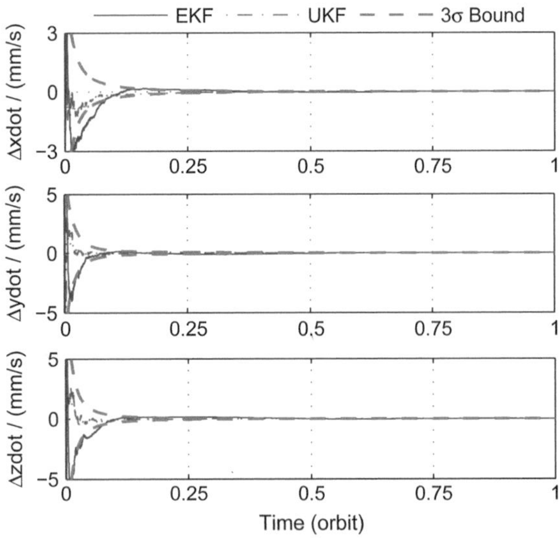

Fig. 3.5 Relative velocity estimation errors. Reprinted from Ref. [1], Copyright 2014, with permission from SAGE

Furthermore, as to the computational considerations, UKF obviates the need to calculate the Jacobians and Hessians, making it easier to implement than EKF. However, the computational complexity of UKF is on the same order as that of EKF, that is, n^3, where n is the dimension of the state vector. Notably, this

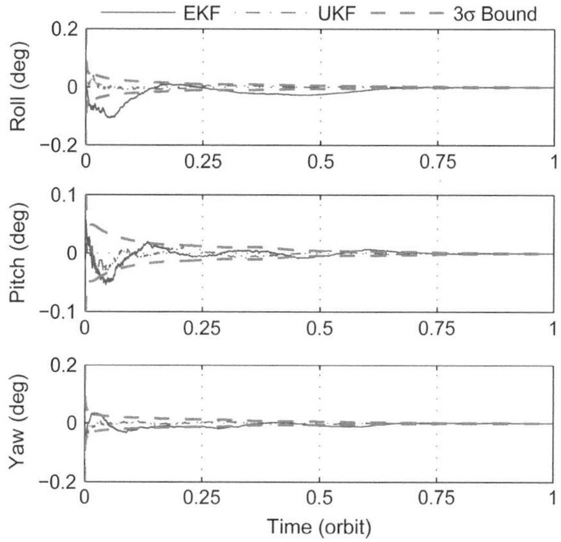

Fig. 3.6 Relative attitude estimation errors. Reprinted from Ref. [1], Copyright 2014, with permission from SAGE

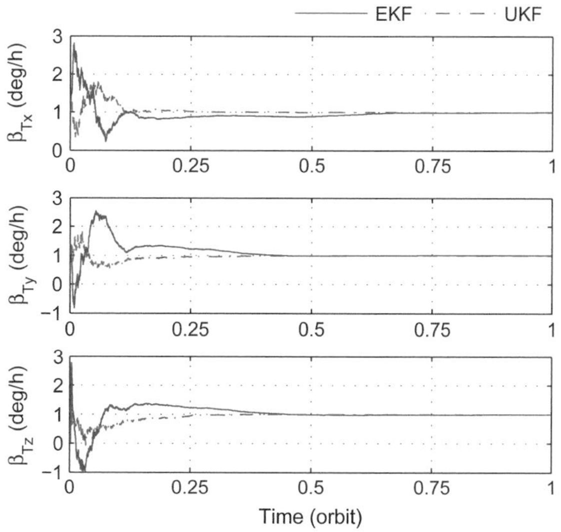

Fig. 3.7 Gyro bias estimate of the target spacecraft. Reprinted from Ref. [1], Copyright 2014, with permission from SAGE

conclusion holds for the original EKF and UKF. Since the state vector is augmented for UKF in this book, then more computational cost and memory are required by UKF as a result of the inclusion of the process and observation noises as states. In other words, if the state vector is not augmented, then the computational cost for both filters are on the same order.

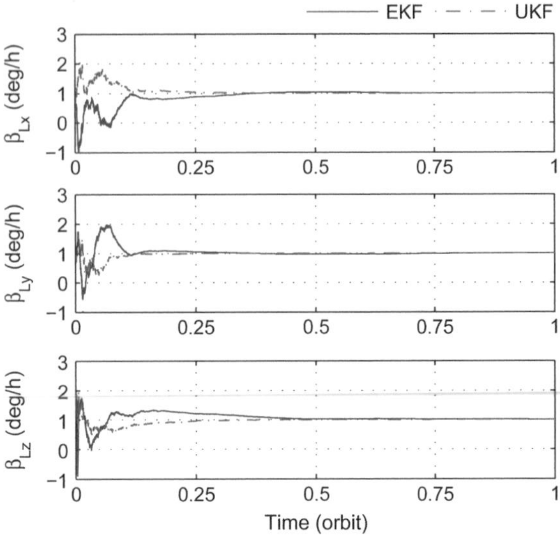

Fig. 3.8 Gyro bias estimate of the Lorentz spacecraft. Reprinted from Ref. [1], Copyright 2014, with permission from SAGE

3.3 Conclusions

Based on the dynamical model of Lorentz-augmented relative orbital motion developed in the last chapter, in combination with the LOS and gyro measurement models, both EKF and UKF algorithms are designed for relative navigation of Lorentz-augmented relative motion. The Lorentz acceleration is determined by the relative position and velocity between spacecraft, which are both states of the dynamical system. Therefore, the incorporation of Lorentz acceleration into system dynamics will further increase the nonlinearity and coupling of relative orbital dynamics. Because EKF is dependent on the linearized dynamical model of relative orbital motion while UKF is dependent on a nonlinear one, then it is no wonder that EKF is inferior to UKF in estimation accuracy. The simulation results indicate that the augmented UKF presents faster convergent rate and superior estimation precision at the expense of more computational cost.

References

1. Huang X, Yan Y, Zhou Y et al (2014) Improved analytical solutions for relative motion of Lorentz spacecraft with application to relative navigation in low Earth orbit. Proc Inst Mech Eng Part G J Aerosp Eng 228:2138–2154. doi:10.1177/0954410013511426
2. Lefferts EJ, Markley FL, Shuster MD (1982) Kalman filtering for spacecraft attitude estimation. J Guid Control Dyn 5:417–429
3. Kim SG, Crassidis JL, Cheng Y, Fosbury AM (2007) Kalman filtering for relative spacecraft attitude and position estimation. J Guid Control Dyn 30:133–143

4. Sun D, Crassidis JL (2002) Observability analysis of six-degree-of-freedom configuration determination using vector observations. J Guid Control Dyn 25:1149–1157
5. Crassidis JL, Markley FL (2003) Unscented filtering for spacecraft attitude estimation. J Guid Control Dyn 26:536–542
6. Haykin S (2001) Kalman filtering and neural networks. Wiley, New York

Chapter 4
Dynamics and Control of Lorentz-Augmented Spacecraft Hovering

Hovering can generally be defined as maintaining a constant position with respect to a target in space. To achieve hovering, a spacecraft thrusts continuously to induce an equilibrium state at a desired position. Due to the constraints on the quantity of propellant onboard, long-time hovering around low-Earth orbits (LEOs) is hardly achievable using traditional chemical propulsion. The Lorentz force, acting on an electrostatically charged spacecraft as it moves through a planetary magnetic field, provides a new propellantless method for orbital maneuvers. This chapter investigates the feasibility of using the induced Lorentz force as an auxiliary means of propulsion for long-term spacecraft hovering.

4.1 Problem Formulation and Dynamical Model

4.1.1 Problem Formulation

On one hand, to maintain a hovering configuration, continuous thrust is generally required to cancel out the relative acceleration between spacecraft. On the other hand, the induced Lorentz force is always perpendicular to the local magnetic field and the vehicle's velocity with respect to the local magnetic field. Due to this constraint, the direction of the Lorentz force does not necessarily coincide with that of the required thrust for hovering. Therefore, in most cases, the Lorentz force works as auxiliary propulsion to reduce the fuel consumption.

In view of this fact, the problem can be formulated as follows.

1. Find the hovering configurations that could achieve propellantless hovering and solve the corresponding required specific charges of a Lorentz spacecraft;
2. For other configurations that necessitate the hybrid inputs consisting of the Lorentz force and chemical propulsion, solve the optimal distribution laws of this two kinds of propulsion to minimize the fuel consumption for hovering;

© Springer Science+Business Media Singapore 2017

Y. Yan et al., *Dynamics and Control of Lorentz-Augmented Spacecraft Relative Motion*, DOI 10.1007/978-981-10-2603-4_4

3. Design closed-loop tracking controllers to track the optimal open-loop control trajectory in the presence of external disturbances and system uncertainties.

4.1.2 Dynamical Model

4.1.2.1 Equations of Relative Motion

As shown in Fig. 4.1, the two spacecraft involved in a hovering configuration are referred to as the chaser and target spacecraft, respectively. The chaser is assumed to be a charged Lorentz spacecraft, but the target is uncharged. $O_E X_I Y_I Z_I$ is an Earth-centered inertial frame, with O_E being the center of Earth. The relative dynamics between spacecraft is described in a local-vertical-local-horizontal (LVLH) frame, denoted by $O_T xyz$, where x axis is along the radial direction of the target, z axis is aligned with the normal direction of the target's orbital plane, and y axis completes the right-handed Cartesian frame. O_T and O_L refer to the center of mass (c.m.) of the target and Lorentz spacecraft, respectively.

Define the position vector of the Lorentz spacecraft with respect to the target as $\boldsymbol{\rho} = \boldsymbol{R}_L - \boldsymbol{R}_T = \begin{bmatrix} x & y & z \end{bmatrix}^{\mathrm{T}}$, where $\boldsymbol{R}_T = \begin{bmatrix} R_T & 0 & 0 \end{bmatrix}^{\mathrm{T}}$ and $\boldsymbol{R}_L = \begin{bmatrix} R_T + x & y & z \end{bmatrix}^{\mathrm{T}}$ are, respectively, the orbital radius of the target and Lorentz spacecraft. Also, define $\boldsymbol{a}_L = \begin{bmatrix} a_x & a_y & a_z \end{bmatrix}^{\mathrm{T}}$ and $\boldsymbol{a}_C = \begin{bmatrix} a_R & a_S & a_W \end{bmatrix}^{\mathrm{T}}$ as the

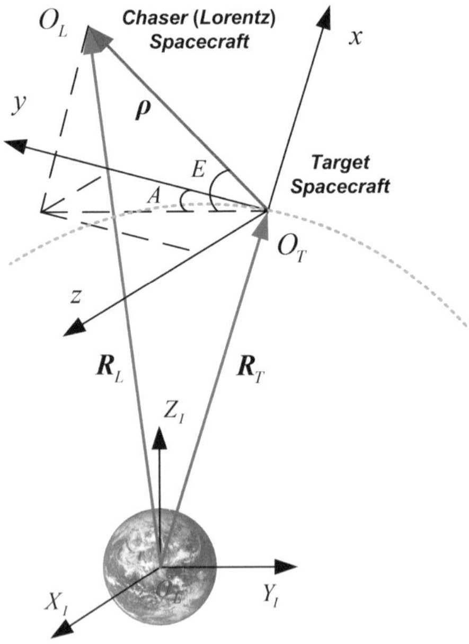

Fig. 4.1 Definitions of the coordinate frames and hovering configuration. Reprinted from Ref. [1], Copyright 2013, with permission from Elsevier

Lorentz acceleration and control acceleration acting on the Lorentz spacecraft, respectively.

Based on the analyses in Sect. 2.2.1.1, the equations of Lorentz-augmented relative motion are given by

$$\begin{aligned}
\ddot{x} &= 2\dot{u}_T\dot{y} + \dot{u}_T^2 x + \ddot{u}_T y + n_T^2 R_T - n_L^2(R_T + x) + a_x + a_R \\
\ddot{y} &= -2\dot{u}_T\dot{x} + \dot{u}_T^2 y - \ddot{u}_T x - n_L^2 y + a_y + a_S \\
\ddot{z} &= -n_L^2 z + a_z + a_W
\end{aligned} \tag{4.1}$$

where $\dot{u}_T$ and $\ddot{u}_T$ refer to the orbital angular velocity and acceleration of the target, respectively. μ is the gravitational parameter of Earth, and the orbital radius of the Lorentz spacecraft is $R_L = [(R_T + x)^2 + y^2 + z^2]^{1/2}$.

Define ρ, E, and A as the hovering parameters, and $\boldsymbol{\Gamma} = [\rho, E, A]$ as the hovering configuration, given by [1]

$$\begin{aligned}
\rho &= \sqrt{x^2 + y^2 + z^2} \\
E &= \arcsin(x/\rho) \\
A &= \arctan(z/y)
\end{aligned} \tag{4.2}$$

where ρ is the hovering distance, $E \in [-\pi/2, \pi/2]$ and $A \in [-\pi, \pi]$ are the hovering elevation and azimuth, respectively. If the chaser hovers above the target, then E is positive. Also if the chaser hovers on the left side of the target, then A is positive. According to the above definitions, the elevation and azimuth in Fig. 4.1 are both positive.

Substitution of Eq. (4.2) into Eq. (4.1) yields the equations of relative motion represented by the polar coordinates [1]:

$$\left\{\begin{aligned}
\ddot{\rho} = {} & [\dot{E}^2 + \dot{A}^2\cos^2 E - \dot{u}_T(2\dot{E}\cos A + \dot{A}\sin 2E\sin A) \\
& + \dot{u}_T^2(\sin^2 E + \cos^2 E\cos^2 A) - \mu/R_L^3]\rho + (R_T^{-2} - R_T R_L^{-3})\mu\sin E \\
& + (a_R + a_x)\sin E + (a_S + a_y)\cos E\cos A + (a_W + a_z)\cos E\sin A \\
\ddot{E} = {} & -(\dot{A}^2\sin 2E)/2 + 2\dot{\rho}(\dot{u}_T\cos A - \dot{E})/\rho - 2\dot{u}_T\dot{A}\cos^2 E\sin A \\
& + (\dot{u}_T^2\sin 2E\sin^2 A)/2 + \ddot{u}_T\cos A + [(R_T^{-2} - R_T R_L^{-3})\mu\cos E \\
& + (a_R + a_x)\cos E - (a_S + a_y)\sin E\cos A - (a_W + a_z)\sin E\sin A]/\rho \\
\ddot{A} = {} & 2\dot{E}\dot{A}\tan E + 2\dot{\rho}(\dot{u}_T\sin A\tan E - \dot{A})/\rho + 2\dot{u}_T\dot{E}\sin A \\
& -(\dot{u}_T^2\sin 2A)/2 + \ddot{u}_T\sin A\tan E - [(a_S + a_y)\sin A \\
& -(a_W + a_z)\cos A]/(\rho\cos E)
\end{aligned}\right. \tag{4.3}$$

4.1.2.2 The Lorentz Acceleration

According to analyses in Sect. 2.2.1.2, the Lorentz acceleration can be expressed in the LVLH frame as

$$\boldsymbol{a}_L = \lambda \boldsymbol{V}_{\text{rel}} \times \boldsymbol{B} = (q_L/m_L)\boldsymbol{V}_{\text{rel}} \times \boldsymbol{B} \tag{4.4}$$

where $\lambda = q_L/m_L$ is the specific charge (i.e., charge-to-mass ratio) of Lorentz spacecraft, $\boldsymbol{V}_{\text{rel}}$ is the velocity of the Lorentz spacecraft relative to the local magnetic field, given by

$$\boldsymbol{V}_{\text{rel}} = \begin{bmatrix} V_x \\ V_y \\ V_z \end{bmatrix} = \begin{bmatrix} \dot{R}_T + \dot{x} - y(\dot{u}_T - \omega_E \cos i_T) - z\omega_E \sin i_T \cos u_T \\ \dot{y} + (R_T + x)(\dot{u}_T - \omega_E \cos i_T) + z\omega_E \sin i_T \sin u_T \\ \dot{z} + (R_T + x)\omega_E \sin i_T \cos u_T - y\omega_E \sin i_T \sin u_T \end{bmatrix} \tag{4.5}$$

where i_T and u_T refer to the orbital inclination and argument of latitude of the target, respectively. ω_E is the rotation rate of Earth.

The magnetic field local at the Lorentz spacecraft is [2]

$$\boldsymbol{B} = \begin{bmatrix} B_x & B_y & B_z \end{bmatrix}^{\mathrm{T}} = (B_0/R_L^3)\left[3(\hat{\boldsymbol{n}} \cdot \hat{\boldsymbol{R}}_L)\hat{\boldsymbol{R}}_L - \hat{\boldsymbol{n}}\right] \tag{4.6}$$

where $\hat{\boldsymbol{R}}_L$ is the unit orbital radius vector of the Lorentz spacecraft. $\hat{\boldsymbol{n}}$ is the unit magnetic dipole momentum vector, which can be expressed in the LVLH frame as

$$\hat{\boldsymbol{n}} = \begin{bmatrix} n_x \\ n_y \\ n_z \end{bmatrix} = \begin{bmatrix} -(\cos\beta \cos u_T + \sin\beta \cos i_T \sin u_T)\sin\alpha - \sin i_T \sin u_T \cos\alpha \\ (\cos\beta \sin u_T - \sin\beta \cos i_T \cos u_T)\sin\alpha - \sin i_T \cos u_T \cos\alpha \\ \sin\beta \sin i_T \sin\alpha - \cos i_T \cos\alpha \end{bmatrix} \tag{4.7}$$

where α is the dipole tilt angle. The angle β is defined as $\beta = \Omega_M - \Omega_T$, where Ω_T is the right ascension of the ascending node of the target, and $\Omega_M = \omega_E t + \Omega_0$ is the inertial phase angle of the magnetic dipole, with t being the time and Ω_0 is the initial phase angle.

If the hovering distance is negligibly small as compared to the orbital radius of either spacecraft (i.e., $\rho \ll R_T, R_L$), the magnetic field local at the Lorentz spacecraft can be approximated by that local the target spacecraft, given by [3]

$$\boldsymbol{B} \approx \frac{B_0}{R_T^3}\begin{bmatrix} -2\sin\alpha(\cos u_T \cos\beta + \sin u_T \cos i_T \sin\beta) - 2\sin u_T \sin i_T \cos\alpha \\ \sin\alpha(-\sin u_T \cos\beta + \cos u_T \cos i_T \sin\beta) + \cos u_T \sin i_T \cos\alpha \\ -\sin\alpha \sin i_T \sin\beta + \cos i_T \cos\alpha \end{bmatrix} \tag{4.8}$$

To analyze the scope of application of this approximation, consider a spherical region with its center located at the target, as shown in Fig. 4.2. The radius of this

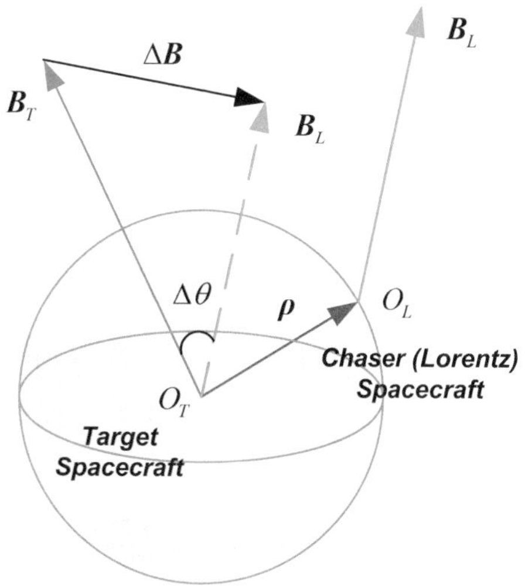

Fig. 4.2 The magnetic field local at the chaser and the target

spherical region (i.e., the relative distance between the chaser and the target) is ρ. Thus, the error magnetic field between the chaser and the target is

$$\Delta \boldsymbol{B} = \boldsymbol{B}_L - \boldsymbol{B}_T \tag{4.9}$$

Define $\Delta\theta$ as the error angle between $\boldsymbol{B}_T$ and $\boldsymbol{B}_L$, and η as the error percentage of geomagnetic field intensity, given by

$$\Delta\theta = \arccos \frac{\boldsymbol{B}_T \cdot \boldsymbol{B}_L}{\|\boldsymbol{B}_T\| \|\mathbf{B}_L\|}, \quad \eta = \frac{\|\Delta \boldsymbol{B}\|}{\|\boldsymbol{B}_T\|} \times 100\,\% \tag{4.10}$$

Using IGRF-11 [4], a numerical assessment of the difference between the two spacecraft in proximity is conducted. Denote $\Delta\theta_{\max}$ (unit: deg) and $\eta_{\max}$ as the maximal error angle and error percentage of geomagnetic field intensity within a relative range of ρ, respectively. The numerical results at different orbital altitudes (i.e., H) are then tabulated in Table 4.1.

According to the numerical simulation results, it can be concluded that the close range assumption is valid within a relative range of tens of kilometers. The maximal error angle and error percentage of geomagnetic field intensity generally decrease with increasing orbital altitudes. For example, to remain the maximal error percentage under 3 % and the error angle less than 1.5° at all orbital altitudes, it is recommended that the hovering distance should be kept within 50 km.

Substitution of Eqs. (4.5) and (4.8) into Eq. (4.4) yields the expressions of Lorentz acceleration in the LVLH frame, given by

Table 4.1 Numerical simulation results

H (km)	ρ (km)					
	10		50		100	
	$\Delta\theta_{max}$	η_{max}	$\Delta\theta_{max}$	η_{max}	$\Delta\theta_{max}$	η_{max}
300	0.26°	0.48 %	1.29°	2.43 %	2.58°	4.94 %
500	0.25°	0.47 %	1.25°	2.36 %	2.50°	4.80 %
1000	0.23°	0.43 %	1.16°	2.20 %	2.33°	4.46 %
2000	0.20°	0.38 %	1.02°	1.93 %	2.05°	3.91 %
10,000	0.10°	0.20 %	0.52°	0.98 %	1.05°	1.98 %
20,000	0.06°	0.12 %	0.32°	0.61 %	0.65°	1.22 %
30,000	0.05°	0.09 %	0.24°	0.44 %	0.47°	0.88 %

$$\boldsymbol{a}_L = \lambda \boldsymbol{l} = \lambda [\, l_x \quad l_y \quad l_z \,]^{\mathrm{T}} \tag{4.11}$$

where

$$l_x = V_y B_z - V_z B_y, \quad l_y = V_z B_x - V_x B_z, \quad l_z = V_x B_y - V_y B_x \tag{4.12}$$

4.2 Case of Two-Body Circular Reference Orbit

4.2.1 Open-Loop Control

According to the definition of hovering, the hovering position remains constant in the LVLH frame, indicating that the time derivative of the hovering position vector is zero, given by

$$\dot{x} = \dot{y} = \dot{z} = 0, \quad \ddot{x} = \ddot{y} = \ddot{z} = 0 \tag{4.13}$$

Substitution of Eq. (4.13) into Eq. (4.1) yields that

$$\begin{aligned} 0 &= \dot{u}_T^2 x + \ddot{u}_T y + n_T^2 R_T - n_L^2 (R_T + x) + a_x + a_R \\ 0 &= \dot{u}_T^2 y - \ddot{u}_T x - n_L^2 y + a_y + a_S \\ 0 &= -n_L^2 z + a_z + a_W \end{aligned} \tag{4.14}$$

Then, the thruster-generated control acceleration $\boldsymbol{a}_C$ can be derived as

$$\boldsymbol{a}_C = \begin{bmatrix} a_R \\ a_S \\ a_W \end{bmatrix} = -\dot{u}_T^2 \begin{bmatrix} x \\ y \\ 0 \end{bmatrix} - \ddot{u}_T \begin{bmatrix} y \\ -x \\ 0 \end{bmatrix} + n_L^2 \begin{bmatrix} R_T + x \\ y \\ z \end{bmatrix} - n_T^2 \begin{bmatrix} R_T \\ 0 \\ 0 \end{bmatrix} - \begin{bmatrix} a_x \\ a_y \\ a_z \end{bmatrix} \tag{4.15}$$

Having derived the general expressions of control acceleration necessary for hovering, detailed analyses will be presented for the following three cases: (1) equatorial circular orbit with nontilted dipole; (2) inclined circular orbit with nontilted dipole; (3) inclined circular orbit with tilted dipole.

4.2.1.1 Equatorial Circular Orbit with Nontilted Dipole

By assuming that the target is flying in an equatorial circular orbit (i.e., $\dot{u}_T = \omega_T = n_T$, $\dot{R}_T = 0$, and $i_T = 0$) and the magnetic dipole is nontilted (i.e., $\alpha = 0$), the magnetic field in Eq. (4.8) reduces to

$$\boldsymbol{B} = (B_0/R_T^3)\hat{\boldsymbol{z}} \tag{4.16}$$

Meanwhile, the relative velocity in Eq. (4.5) reduces to

$$\boldsymbol{V}_{\text{rel}} = [-y(\omega_T - \omega_E)]\hat{\boldsymbol{x}} + [(R_T + x)(\omega_T - \omega_E)]\hat{\boldsymbol{y}} \tag{4.17}$$

where ω_T refers to the orbital angular velocity of the target, and the superscript ^ denotes the unit vector on the axis.

Substitution of Eqs. (4.16) and (4.17) into Eq. (4.11) yields the expression of Lorentz acceleration, given by

$$\boldsymbol{a}_L = \lambda\boldsymbol{l} = \lambda(B_0/R_T^3)(\omega_T - \omega_E)[(R_T + x)\hat{\boldsymbol{x}} + y\hat{\boldsymbol{y}}] \tag{4.18}$$

Thus, the thruster-generated control acceleration can be derived via Eq. (4.15) as

$$\boldsymbol{a}_C = \boldsymbol{h} - \lambda\boldsymbol{l} = \begin{bmatrix} (n_L^2 - n_T^2)(R_T + x) - \lambda(B_0/R_T^3)(\omega_T - \omega_E)(R_T + x) \\ (n_L^2 - n_T^2)y - \lambda(B_0/R_T^3)(\omega_T - \omega_E)y \\ n_L^2 z \end{bmatrix} \tag{4.19}$$

where $\boldsymbol{h}$ is the total acceleration necessary for hovering, given by

$$\boldsymbol{h} = \begin{bmatrix} h_x \\ h_y \\ h_z \end{bmatrix} = \begin{bmatrix} (n_L^2 - n_T^2)(R_T + x) \\ (n_L^2 - n_T^2)y \\ n_L^2 z \end{bmatrix} \tag{4.20}$$

As can be seen, for this case, the Lorentz acceleration only acts in the orbital plane, and has no effect on the out-of-plane relative motion. It indicates that the normal control acceleration will be fully provided by the thrusters. Thus, to minimize the fuel consumption for hovering, it is desirable to find the specific charge of Lorentz spacecraft that renders

$$a_R = a_S = 0 \tag{4.21}$$

In view of Eq. (4.19), the solution to Eq. (4.21) is

$$\lambda = \frac{q_L}{m_L} = \frac{n_L^2 - n_T^2}{\omega_T - \omega_E} \frac{R_T^3}{B_0} \tag{4.22}$$

From Eq. (4.22), following conclusions can be derived:

1. The polarity of the specific charge is dependent on the orbital radius of the Lorentz spacecraft and the target spacecraft;
2. Note that $R_L^2 = R_T^2 + \rho^2 + 2\rho R_T \sin E$, the required specific charge for this case is only determined by two of the three hovering parameters, that is, ρ and E;
3. If the Lorentz spacecraft hovers in the target's equatorial plane that the normal hovering position is zero (i.e., $z = 0$), then no thruster-generated control acceleration is required and a propellantless hovering can be achieved by only using the Lorentz force;
4. At the altitude of geostationary orbit (GEO) where $\omega_T = \omega_E$, the Lorentz spacecraft has no relative velocity with respect to the local magnetic field and no Lorentz force can thus be induced for hovering.

Define a critical elevation E_C that satisfies $R_L = R_T$, given by

$$E_C = \arcsin\left[-\rho/(2R_T)\right] \tag{4.23}$$

Note that $\rho \ll R_T$, then it holds that $E_C \approx 0$. Also, if $E > E_C$, then it holds that $R_L > R_T$ and thus $n_L < n_T$. Using this relationship, it can be obtained that for orbits below GEO where $\omega_T > \omega_E$, it has

$$\begin{aligned} \lambda &= -|q_L/m_L|, E > E_C \\ \lambda &= +|q_L/m_L|, E < E_C \end{aligned} \tag{4.24}$$

Also, for orbits above GEO where $\omega_T < \omega_E$, it has

$$\begin{aligned} \lambda &= -|q_L/m_L|, E < E_C \\ \lambda &= +|q_L/m_L|, E > E_C \end{aligned} \tag{4.25}$$

Furthermore, define T as the orbital period of the target. Then, for a Lorentz spacecraft with Lorentz force as auxiliary propulsion for hovering, the velocity increment consumption for one-orbital-period hovering will be

$$\Delta V = \int_0^T ||\boldsymbol{a}_C|| \mathrm{d}t = a_W T \tag{4.26}$$

Differently, for an uncharged spacecraft that the required control acceleration for hovering is fully provided by the thrusters on board, the consumption of velocity increment is

$$\Delta V = \int_0^{T} ||\boldsymbol{h}||\mathrm{d}t = ||\boldsymbol{h}||T \tag{4.27}$$

4.2.1.2 Inclined Circular Orbit with Nontilted Dipole

For this case, the target's orbit is an inclined circular one (i.e., $i_T \neq 0$). Also the magnetic dipole is assumed to be nontilted. Then the magnetic field for this case is reduced to

$$\boldsymbol{B} = \frac{B_0}{R_T^3}\begin{bmatrix} -2\sin u_T \sin i_T \\ \cos u_T \sin i_T \\ \cos i_T \end{bmatrix} \tag{4.28}$$

Also, the relative velocity for this case is

$$\boldsymbol{V}_{\mathrm{rel}} = \begin{bmatrix} -y(\omega_T - \omega_E \cos i_T) - z\omega_E \sin i_T \cos u_T \\ (R_T + x)(\omega_T - \omega_E \cos i_T) + z\omega_E \sin i_T \sin u_T \\ (R_T + x)\omega_E \sin i_T \cos u_T - y\omega_E \sin i_T \sin u_T \end{bmatrix} \tag{4.29}$$

where $u_T = u_{T0} + \omega_T t$, with u_{T0} being the initial argument of latitude.

Likewise, substitution of Eqs. (4.28) and (4.29) into Eq. (4.11) yields the Lorentz acceleration for this case, given by

$$\boldsymbol{a}_L = \lambda[l_x(u_T)\hat{\boldsymbol{x}} + l_y(u_T)\hat{\boldsymbol{y}} + l_z(u_T)\hat{\boldsymbol{z}}] \tag{4.30}$$

where

$$\begin{aligned} l_x &= (B_0/R_T^3)\{(R_T + x)\left[(\omega_T - \omega_E \cos i_T)\cos i_T - \omega_E \sin^2 i_T \cos^2 u_T\right] \\ &\quad + (y \sin i_T \cos u_T + z \cos i_T)\omega_E \sin i_T \sin u_T\} \\ l_y &= (B_0/R_T^3)\{-(R_T + x)\omega_E \sin^2 i_T \sin 2u_T + y[(\omega_T - \omega_E \cos i_T)\cos i_T \\ &\quad + 2\omega_E \sin^2 i_T \sin^2 u_T] + z\omega_E \sin i_T \cos i_T \cos u_T\} \\ l_z &= (B_0/R_T^3)\{[2(R_T + x)\sin u_T - y \cos u_T](\omega_T - \omega_E \cos i_T)\sin i_T \\ &\quad + z\omega_E \sin^2 i_T(3\sin^2 u_T - 1)\} \end{aligned} \tag{4.31}$$

Thus, the thruster-generated control acceleration necessary for hovering is

$$\boldsymbol{a}_C = \boldsymbol{h} - \lambda \boldsymbol{l} \tag{4.32}$$

As shown in Eq. (4.31), $\boldsymbol{l}$ is time-varying and it determines the direction of the Lorentz acceleration. Meanwhile, it indicates that once the argument of latitude u_T is determined, the direction of the Lorentz acceleration $\boldsymbol{l}$ is also determined. If the direction of Lorentz acceleration is parallel to that of the total acceleration for hovering (i.e., $\boldsymbol{l} \parallel \boldsymbol{h}$), then by setting the specific charge of the Lorentz spacecraft as $\lambda = ||\boldsymbol{h}||/||\boldsymbol{l}||$, the resulting Lorentz acceleration can fully compensate for the required total acceleration for hovering. However, $\boldsymbol{h}$ is time-invariant but $\boldsymbol{l}$ is time-varying. Therefore, there is no guarantee that the direction of the induced Lorentz acceleration is always parallel to that of the total acceleration for hovering. Then, other kinds of propulsion such as the thruster-generated control acceleration are also required to achieve hovering. Actually, in most cases, the Lorentz acceleration works as auxiliary propulsion to reduce the fuel consumptions. Given that the propellant is a limited resource on board, an energy-optimal objective function in Lagrange form is selected to minimize the energy consumption for hovering, given by

$$J = \int_0^{\mathrm{T}} L[t, \lambda(t)]\mathrm{d}t = \frac{1}{2}\int_0^{\mathrm{T}} \boldsymbol{a}_C^{\mathrm{T}} \boldsymbol{a}_C \mathrm{d}t \tag{4.33}$$

Solving the Euler–Lagrange equation

$$\frac{\mathrm{d}}{\mathrm{d}t}\left(\frac{\partial L}{\partial \dot{\lambda}}\right) - \frac{\partial L}{\partial \lambda} = 0 \tag{4.34}$$

yields the energy-optimal trajectory of the specific charge of Lorentz spacecraft, given by

$$\lambda^*(t) = \begin{cases} (\boldsymbol{h} \cdot \boldsymbol{l})/||\boldsymbol{l}||^2, & ||\boldsymbol{l}|| \neq 0 \\ 0, & ||\boldsymbol{l}|| = 0 \end{cases} \tag{4.35}$$

Thus, the energy-optimal trajectories of the thruster-generated control acceleration and the Lorentz acceleration are, respectively, derived as

$$\boldsymbol{a}_C^*(t) = \begin{cases} \boldsymbol{h} - [(\boldsymbol{h} \cdot \boldsymbol{l})/||\boldsymbol{l}||^2]\boldsymbol{l}, & ||\boldsymbol{l}|| \neq 0 \\ \boldsymbol{h}, & ||\boldsymbol{l}|| = 0 \end{cases} \tag{4.36}$$

and

$$\boldsymbol{a}_L^*(t) = \begin{cases} [(\boldsymbol{h} \cdot \boldsymbol{l})/||\boldsymbol{l}||^2]\boldsymbol{l}, & ||\boldsymbol{l}|| \neq 0 \\ \boldsymbol{0}, & ||\boldsymbol{l}|| = 0 \end{cases} \tag{4.37}$$

In view of Eqs. (4.36) and (4.37), it is notable that $\boldsymbol{a}_C^* \cdot \boldsymbol{a}_L^* = 0$, indicating that the optimal thruster-generated control acceleration is always perpendicular to the Lorentz acceleration. The physical intuition for this fact can be expressed by the geometric relationship shown in Fig. 4.3. At a given time t, the total acceleration necessary for hovering is $\boldsymbol{h}$, and the direction of Lorentz acceleration is determined by $\boldsymbol{l}[u_T(t)]$. If the charge is positive $\lambda > 0$, then the direction of Lorentz acceleration $\lambda\boldsymbol{l}$ is the same as that of $\boldsymbol{l}$. Otherwise, when $\lambda < 0$, the Lorentz acceleration acts in an opposite direction of $\boldsymbol{l}$. Therefore, the specific charge λ is the only variable that can be altered to change the magnitude of Lorentz acceleration. Once λ is determined, $\lambda\boldsymbol{l}$ is then determined, and the thruster-generated control acceleration $\boldsymbol{a}_C$ can thereafter be determined by completing the decomposition of $\boldsymbol{h}$. Obviously, $\boldsymbol{a}_C$ reaches its minimum when perpendicular to $\lambda\boldsymbol{l}$. Using this geometric relationship, it can also be derived that the optimal specific charge of Lorentz spacecraft is $\lambda^* = (\boldsymbol{h} \cdot \boldsymbol{l})/||\boldsymbol{l}||^2$, verifying the validity of Eq. (4.35). Furthermore, when the direction of the relative velocity $\boldsymbol{V}_{\text{rel}}$ is parallel to that of the magnetic field $\boldsymbol{B}$, no Lorentz acceleration will be induced because $\boldsymbol{l} = \boldsymbol{V}_{\text{rel}} \times \boldsymbol{B} = 0$. If this is the case, the total acceleration will fully be provided by the thrusters on board and the specific charge λ is thus set as zero, which is also consistent with the analytic result in Eq. (4.35).

Likewise, for a charged Lorentz spacecraft, the velocity increment for hovering during one-orbital period is

$$\Delta V = \int_0^T ||\boldsymbol{a}_C^*||\mathrm{d}t \tag{4.38}$$

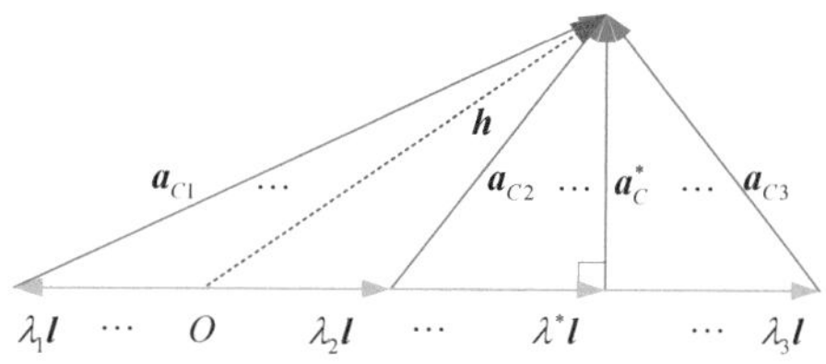

Fig. 4.3 Illustration of optimal thruster-generated control acceleration. Reprinted from Ref. [1], Copyright 2013, with permission from Elsevier

Substitution of Eq. (4.36) into Eq. (4.33) yields the minimal energy consumption, given by

$$J^* = \frac{1}{2}\int_0^T ||\boldsymbol{a}_C^*||^2 \mathrm{d}t = \frac{1}{2}\int_0^T (||\boldsymbol{h}||^2 - ||\boldsymbol{a}_L^*||^2)\mathrm{d}t \tag{4.39}$$

As can be seen, at a given reference orbital altitude R_T, J^* is the function of the reference orbital inclination i_T. To investigate the effect of i_T on J^*, define a critical inclination i_T^c that satisfies

$$J^*(i_T^c) = \max J^*(i_T) \tag{4.40}$$

It indicates that at the critical inclination i_T^c, J^* reaches its maximum and the induced Lorentz acceleration provides least efficiency in reducing the control energy consumption. The critical inclination for each orbital altitude can simply be derived by numerical methods such as one-dimensional searching method.

Also, an approximate analytical expression of the critical inclination is derived in the following theorem. Notably, this expressions is valid for orbits below GEO where $\omega_T > \omega_E$, and the orbits near GEO are also neglected due to the ineffectiveness of Lorentz spacecraft in such orbits. Specifically, the orbital altitudes discussed here mainly range from 300 to 30,000 km.

Theorem 4.1 *For orbits below GEO, if the chaser hovers in the radial direction of the target (i.e.,* $\boldsymbol{\Gamma} = [\rho, \pm\pi/2, 0]$*), the critical inclination can be approximated by*

$$i_T^c \approx \arccos(\omega_E/\omega_T) \tag{4.41}$$

Also, Eq. (4.41) *holds when the hovering distance is negligibly small as compared to the orbital radius of either spacecraft.*

Proof Define a coefficient as $\gamma = \omega_E/\omega_T$. For orbits below GEO, it holds that $0<\gamma<1$. If the chaser hovers in the radial direction of the target that $y = z = 0$, then the minimal control energy consumption can be represented as

$$\begin{aligned} J^* &= \frac{1}{2}\int_0^T \left[h_x^2 - (h_x l_x)^2/(l_x^2 + l_y^2 + l_z^2)\right]\mathrm{d}t \\ &= \frac{1}{2\omega_T}\int_0^{2\pi} \left[h_x^2 - (h_x l_x)^2/(l_x^2 + l_y^2 + l_z^2)\right]\mathrm{d}u_T \end{aligned} \tag{4.42}$$

where

$$\begin{aligned} l_x &= (R_T + x)\left[(\omega_T \cos i_T - \omega_E) + \omega_E \sin^2 i_T \sin^2 u_T\right] \\ l_y &= -(R_T + x)\omega_E \sin^2 i_T \sin^2 u_T \\ l_z &= 2(R_T + x)(\omega_T - \omega_E \cos i_T) \sin i_T \sin u_T \end{aligned} \tag{4.43}$$

Notably, Eq. (4.42) also holds when the hovering distance is negligibly small as compared to the orbital radius of either spacecraft.

Define $\tilde{J}^*$ as the normalized optimal energy consumption, given by

$$\tilde{J}^* = \frac{2\omega_T J^*}{h_x^2} = \int_0^{2\pi} \tilde{L} \mathrm{d}u_T \tag{4.44}$$

where

$$\tilde{L} = 1 - \frac{l_x^2}{l_x^2 + l_y^2 + l_z^2} \tag{4.45}$$

Substitution of Eq. (4.42) into Eq. (4.44) yields that

$$\tilde{L} = 1 - \frac{a_1 \sin^4 u_T + b_1 \sin^2 u_T + c_1}{a_2 \sin^4 u_T + b_2 \sin^2 u_T + c_2} \tag{4.46}$$

where

$$\begin{aligned} a_1 &= \omega_E^2 \sin^4 i_T \\ b_1 &= 2(\omega_T \cos i_T - \omega_E)\omega_E \sin^2 i_T \\ c_1 &= (\omega_T \cos i_T - \omega_E)^2 \\ a_2 &= -3\omega_E^2 \sin^4 i_T = -3a_1 \\ b_2 &= 2(2\omega_T^2 - 3\omega_T\omega_E \cos i_T + \omega_E^2) \sin^2 i_T \\ c_2 &= (\omega_T \cos i_T - \omega_E)^2 = c_1 \end{aligned} \tag{4.47}$$

When $c_2 \neq 0$, that is, $\cos i_T \neq \gamma$, Eq. (4.46) reduces to

$$
\begin{aligned}
\tilde{L} &= \frac{4}{3} - \frac{(b_1 + b_2/3)\sin^2 u_T}{a_2(\sin^2 u_T + \lambda_1)(\sin^2 u_T - \lambda_2)} - \frac{c_1 + c_2/3}{a_2(\sin^2 u_T + \lambda_1)(\sin^2 u_T - \lambda_2)} \\
&= \frac{4}{3} - \frac{3b_1 + b_2}{3a_2(\lambda_1 + \lambda_2)}\left[\frac{\lambda_1}{\lambda_1 \cos^2 u_T + (\lambda_1 + 1)\sin^2 u_T} - \frac{\lambda_2}{\lambda_2 \cos^2 u_T + (\lambda_2 - 1)\sin^2 u_T}\right] \\
&\quad + \left[\frac{1}{\lambda_1 \cos^2 u_T + (\lambda_1 + 1)\sin^2 u_T} + \frac{1}{\lambda_2 \cos^2 u_T + (\lambda_2 - 1)\sin^2 u_T}\right]\frac{4c_1}{3a_2(\lambda_1 + \lambda_2)}
\end{aligned}
\tag{4.48}
$$

where

$$
\lambda_1 = \frac{b_2 - \sqrt{b_2^2 - 4a_2c_2}}{2a_2}, \quad \lambda_2 = \frac{-b_2 - \sqrt{b_2^2 - 4a_2c_2}}{2a_2} \tag{4.49}
$$

It can be proved that when $0 < \gamma < 1$, it holds that $\lambda_1 > 0$ and $\lambda_2 > 1$. Substitution of Eq. (4.48) into Eq. (4.44) yields that

$$
\tilde{J}^* = \frac{8\pi}{3} - \frac{2\pi}{3a_2(\lambda_1 + \lambda_2)}\left[\frac{(3b_1 + b_2)\lambda_1 - 4c_1}{\sqrt{\lambda_1(\lambda_1 + 1)}} - \frac{(3b_1 + b_2)\lambda_2 + 4c_1}{\sqrt{\lambda_2(\lambda_2 - 1)}}\right] \tag{4.50}
$$

Note that when $c_2 = 0$, that is, $\cos i_T = \gamma = \cos i_T^c$, it holds that $\lambda_1 = 0$ and $\lambda_2 = -b_2/a_2$. If this is the case, Eq. (4.50) is not tenable, and $\tilde{L}$ and $\tilde{J}^*$ can be rewritten as

$$
\tilde{L} = \frac{4}{3} - \frac{1}{3} \cdot \frac{1}{\cos^2 u_T + (1 + a_2/b_2)\sin^2 u_T} \tag{4.51}
$$

and

$$
\tilde{J}^*(i_T^c) = \frac{8\pi}{3} - \frac{2\pi}{3\sqrt{1 - 3\gamma^2/4}} \tag{4.52}
$$

Next, it will prove the continuity of $\tilde{J}^*$ at the switch point $i_T = i_T^c$. Note that

$$
\begin{aligned}
\lim_{i_T \to i_T^c} \tilde{J}^* &= \lim_{i_T \to i_T^c} \int_0^{2\pi} \tilde{L} du_T \\
&= \lim_{i_T \to i_T^c} \int_0^{2\pi} \left[\frac{4}{3} - \frac{1}{3} \frac{\lambda_2}{\lambda_2 \cos^2 u_T + (\lambda_2 - 1) \sin^2 u_T} \right] du_T \\
&= \lim_{i_T \to i_T^c} \left[\frac{8\pi}{3} - \frac{1}{3} \frac{2\pi\lambda_2}{\sqrt{\lambda_2(\lambda_2 - 1)}} \right] \\
&= \frac{8\pi}{3} - \frac{2\pi}{3\sqrt{1 - 3\gamma^2/4}} = \tilde{J}^*(i_T^c)
\end{aligned}
\tag{4.53}
$$

Then, from Eq. (4.53), it is easy to obtain that $\tilde{J}^*$ is continuous at the switch point $i_T = i_T^c$. Thus, $\tilde{J}^*$ can be rewritten as

$$
\tilde{J}^* = \begin{cases} \frac{8\pi}{3} - \frac{2\pi}{3a_2(\lambda_1 + \lambda_2)} \left[\frac{(3b_1 + b_2)\lambda_1 - 4c_1}{\sqrt{\lambda_1(\lambda_1 + 1)}} - \frac{(3b_1 + b_2)\lambda_2 + 4c_1}{\sqrt{\lambda_2(\lambda_2 - 1)}} \right], & i_T \neq i_T^c \\ \frac{8\pi}{3} - \frac{2\pi}{3\sqrt{1 - 3\gamma^2/4}}, & i_T = i_T^c \end{cases}
\tag{4.54}
$$

As aforementioned, the orbital altitudes discussed here range from 300 to 30,000 km, that is, $0.063 < \gamma < 0.799$. Using Eq. (4.54), it can be proved that within this range, it holds that $\tilde{J}^*(i_T) < \tilde{J}^*(i_T^c)$. In other words, $\tilde{J}^*$ reaches its maximum at the critical inclination i_T^c. Detailed mathematical proof is not given here for brevity, but typical trajectories of $\tilde{J}^*$ with varying inclinations at different altitudes are shown in Fig. 4.4 to substantiate the argument. Notably, in Fig. 4.4, from the left to the right, each line represents the orbital altitude of 30,000, 20,000, 10,000, 5000, and 300 km, respectively. As can be seen, for all the displayed orbital altitudes, $\tilde{J}^*$ reaches its maximum at the critical inclination i_T^c. This completes the proof.

4.2.1.3 Inclined Circular Orbit with Tilted Dipole

For this case, the target's orbit is an inclined circular one (i.e., $i_T \neq 0$), and the magnetic dipole is tilted (i.e., $\alpha \neq 0$).

Likewise, the relative velocity for this case is shown in Eq. (4.29), and the magnetic field for this case is shown in Eq. (4.8). The only difference with the case of inclined circular orbit with nontilted dipole (i.e., Sect. 4.2.1.2) is the inclusion of the dipole tilt angle, resulting in the difference of $\boldsymbol{B}$ and thus $\boldsymbol{l}$. Thus, by using similar methods in Sect. 4.2.1.2, it can be derived that the energy-optimal trajectory of the specific charge of the Lorentz spacecraft is also Eq. (4.35). Correspondingly, the optimal trajectories of the thruster-generated control acceleration and the Lorentz acceleration are also given in Eq. (4.36) and Eq. (4.37), respectively.

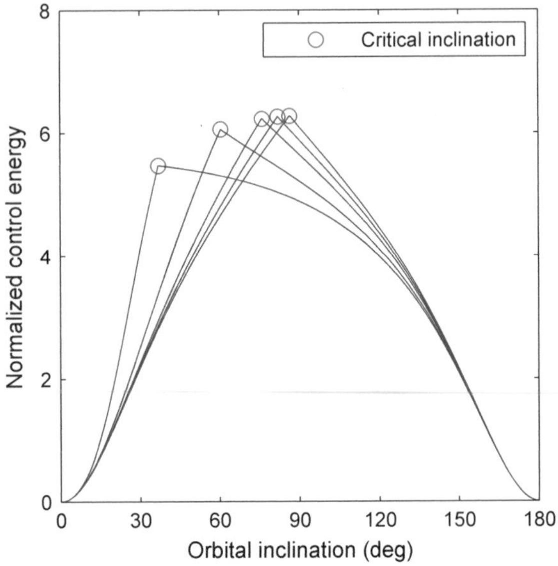

Fig. 4.4 Normalized control energy with varying inclinations at different orbital altitudes. Reprinted from Ref. [1], Copyright 2013, with permission from Elsevier

4.2.2 *Closed-Loop Control*

In the following Sect. 4.3.2, the closed-loop control schemes for Lorentz-augmented spacecraft hovering around elliptic reference orbits will be detailedly introduced. Given that the closed-loop controllers for elliptic reference orbits are also applicable to circular ones, discussions for circular reference orbits are thus not elaborated here separately. Detailed closed-loop controller design is referred to Sect. 4.3.2.

4.2.3 *Numerical Simulation*

Numerical simulations are presented in this section to verify the validity of the proposed energy-optimal open-loop control schemes for the following three cases.

4.2.3.1 Equatorial Circular Orbit with Nontilted Dipole

The target is assumed to be flying in an equatorial circular orbit with an orbital altitude of 500 km. As discussed in Sect. 4.2.1.1, for this case, the specific charge of Lorentz spacecraft necessary for hovering is irrelevant to the hovering azimuth. Therefore, the required specific charges for different pairs of hovering distance and elevation at this orbital altitude are shown in Fig. 4.5. For example, if the chaser

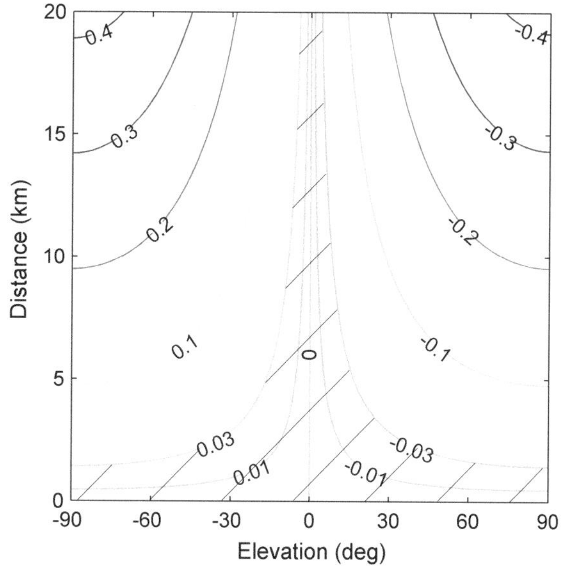

Fig. 4.5 Required specific charges for different hovering distances and elevations. Reprinted from Ref. [1], Copyright 2013, with permission from Elsevier

hovers radially above the target by 5 km (i.e., $\rho = 5$ km and $E = 90°$), the necessary specific charge is about -0.1 C/kg. As shown in Fig. 4.5, the specific charge necessary for hovering generally increases with increasing hovering distance. Furthermore, for hovering in close proximity, the critical elevation E_C is about zero. When $E < E_C$, the necessary specific charge is positive, and when $E > E_C$, the necessary specific charge is negative, verifying the validity of Eq. (4.24). Given that the near-term feasible maximal specific charge is about ± 0.03 C/kg, then the corresponding near-term feasible hovering configurations are shown by the shaded areas in Fig. 4.5.

For a charged Lorentz spacecraft, the velocity increment consumptions ΔV (unit: m/s) for one-orbit hovering at different hovering elevations and azimuths are depicted in Fig. 4.6. Also, those of an uncharged spacecraft are shown in Fig. 4.7. In both figures, the hovering distance is set as 5 km. As shown in Fig. 4.6, for a charged Lorentz spacecraft, only the normal thruster-generated control acceleration a_W is required to achieve hovering, and a_W is proportional to the normal hovering position, that is, $z = \rho \cos E \sin A$. Thus, the required thruster-generated velocity increments for a hovering Lorentz spacecraft are centrosymmetric. Furthermore, if the Lorentz spacecraft hovers in the target's orbital plane that $A = 0$, then the Lorentz acceleration could fully compensate for the total acceleration for hovering, and no thruster-generated control acceleration is required. Contrarily, for hovering in the normal direction of the target that $A = \pm 90°$, the consumption of velocity increment is typically high as compared to other azimuths because the Lorentz acceleration cannot act in the normal direction for this case.

As compared to those of an uncharged spacecraft as shown in Fig. 4.7, it can be found that the consumptions of velocity increments could be greatly reduced using

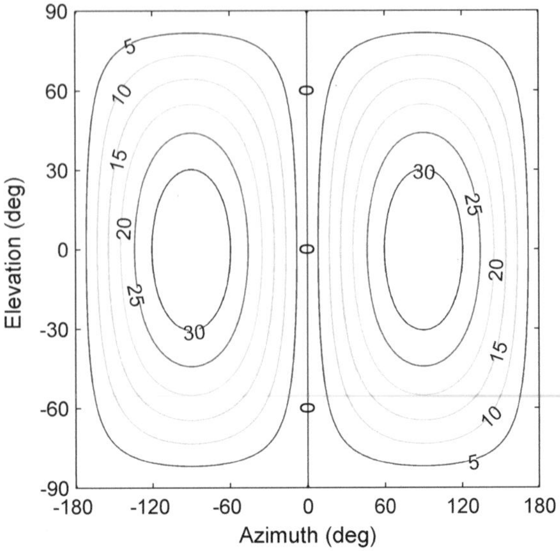

Fig. 4.6 Required velocity increments for a hovering Lorentz spacecraft (*unit* m/s). Reprinted from Ref. [1], Copyright 2013, with permission from Elsevier

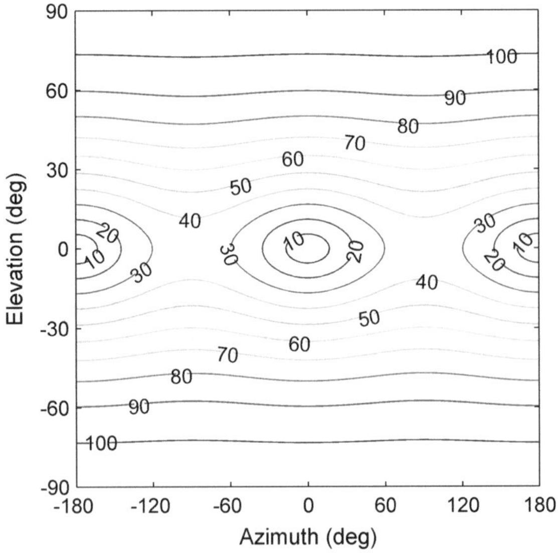

Fig. 4.7 Required velocity increments for a hovering uncharged spacecraft (*unit* m/s). Reprinted from Ref. [1], Copyright 2013, with permission from Elsevier

the Lorentz force as auxiliary propulsion, especially for hovering configurations mainly in the orbital plane, that is, $E \to 90°$ or $E \to -90°$. Take the hovering configuration $\mathbf{\Gamma} = [5\ \text{km}, \pi/3, 2\pi/3]$, for example. For a charged Lorentz spacecraft, the velocity increment for one-orbit hovering is about 15 m/s. However, for an uncharged spacecraft, it increases to 90 m/s, which is about six times that of the Lorentz spacecraft. Likewise, by substituting Eqs. (4.2) and (4.20) into Eq. (4.27),

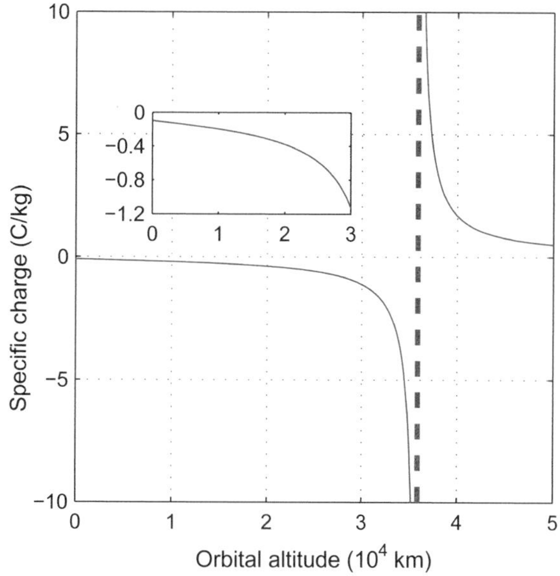

Fig. 4.8 Required specific charge at different orbital altitude (**Γ** = [5 km, π/2, 0]). Reprinted from Ref. [1], Copyright 2013, with permission from Elsevier

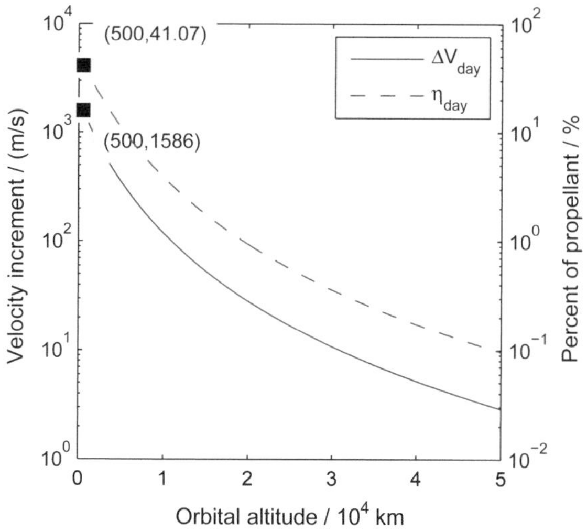

Fig. 4.9 Required specific charge and percent of propellant of an uncharged spacecraft for one-day hovering at different orbital altitudes (**Γ** = [5 km, π/2, 0]). Reprinted from Ref. [1], Copyright 2013, with permission from Elsevier

it can be obtained that the required thruster-generated velocity increments for a hovering uncharged spacecraft are also centrosymmetric.

To analyze the effect of the target's orbital altitude on the charging levels for hovering, the necessary specific charges for keeping the hovering configuration **Γ** = [5 km, π/2, 0] at different orbital altitudes are shown in Fig. 4.8. Note that for this hovering configuration where the chaser hovers in the target's orbital plane, no thruster-generated control acceleration is required for a Lorentz spacecraft and a propellantless hovering can be achieved using the Lorentz force only. As can be seen, because the Lorentz spacecraft is less effective in orbits near GEO, the required specific charges for hovering increase significantly for orbits near GEO. Furthermore, given that $E = \pi/2 > E_C$, the specific charge is negative for orbits below GEO, and is positive for orbits above GEO, verifying the validity of Eqs. (4.24) and (4.25) again. Figure 4.9 shows the necessary velocity increment consumption for keeping this hovering configuration in one day (24 h) ΔV_{day} at different altitudes. Meanwhile, the corresponding necessary percent of the mass of propellant in the total mass of spacecraft η_{day} is also depicted in Fig. 4.9. The percent of the consumed propellant is calculated via

$$\eta_{\text{day}} = 1 - e^{-\Delta V_{\text{day}}/u_r} \tag{4.55}$$

where u_r is the specific impulse of the thruster, which is assumed to be 3 km/s in this chapter.

As shown in Fig. 4.9, for an uncharged spacecraft, much more propellant is required to maintain this configuration in LEOs than high Earth orbits. For example, to maintain the hovering configuration $\boldsymbol{\Gamma}$ = [5 km, π/2, 0] in one day at the altitude of 500 km, the necessary velocity increment increases to 1.586 km/s, which corresponds to a 41.07 % consumption of the total mass of the spacecraft. Such a huge consumption makes the long-term hovering hardly achievable in LEOs by using traditional chemical propulsion. Contrarily, as analyzed above, a charged Lorentz spacecraft can maintain this configuration with no propellant cost. Therefore, if a charged Lorentz spacecraft is used, then no propellant is required (i.e., $\Delta V_{\text{day}} = 0$ and $\eta_{\text{day}} = 0$), making long-term hovering in LEOs feasible.

4.2.3.2 Inclined Circular Orbit with Nontilted Dipole

The initial orbit elements of the target spacecraft are given in Table 4.2, and the hovering configuration is set as $\boldsymbol{\Gamma}$ = [1 km, $-\pi$/3, $-\pi$/4]. The energy-optimal trajectory of the specific charge of Lorentz spacecraft is shown in Fig. 4.10, and the energy-optimal trajectories of the thruster-generated control acceleration for both the Lorentz spacecraft and the uncharged spacecraft are compared in Fig. 4.11. As can be seen, with the Lorentz force as auxiliary propulsion for hovering, the thruster-generated control acceleration could be effectively reduced. For an

Table 4.2 Initial orbit elements of the target

Orbit element	Value
Semi-major axis (km)	6900
Eccentricity	0
Inclination (deg)	15
Right ascension of ascending node (deg)	50
Argument of latitude (deg)	0

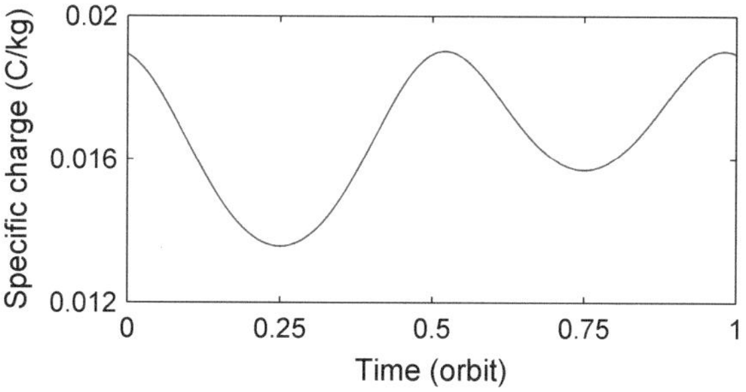

Fig. 4.10 Energy-optimal trajectory of specific charge for hovering

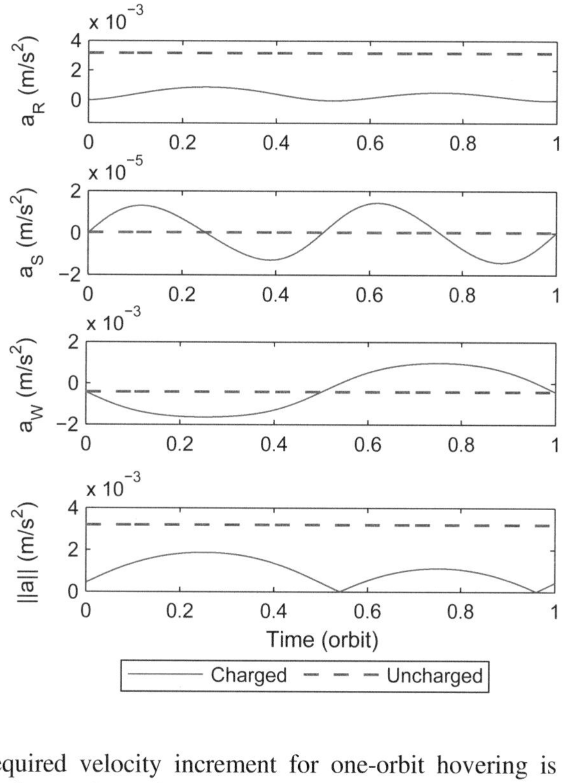

Fig. 4.11 Energy-optimal trajectory of thruster-generated control acceleration for hovering

uncharged spacecraft, the required velocity increment for one-orbit hovering is about 18.15 m/s. However, for a charged Lorentz spacecraft, it decreases to 5.84 m/s and nearly 67.83 % of the velocity increment could be saved, verifying the advantage of Lorentz spacecraft in saving fuels.

Figure 4.12 shows the optimal control energy function at different inclinations for this hovering configuration. As can be seen, the control energy function reaches its maximum at the inclination of about 86°, which is coincided with the approximated analytical solution calculated via Eq. (4.41).

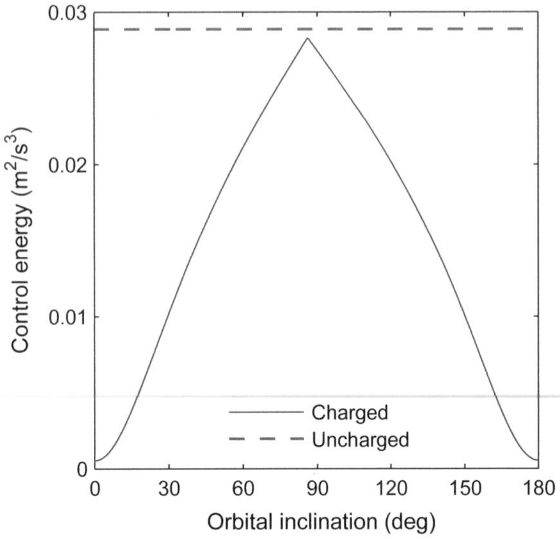

Fig. 4.12 Optimal control energy at different inclinations ($\mathbf{\Gamma}$ = [1 km, $-\pi/3$, $-\pi/4$])

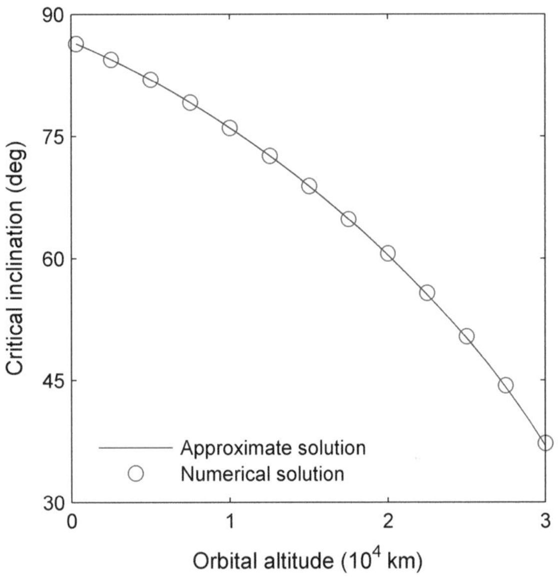

Fig. 4.13 Critical inclinations at different orbital altitudes. Reprinted from Ref. [1], Copyright 2013, with permission from Elsevier

Furthermore, Fig. 4.13 shows the critical inclinations at different orbital altitudes. Obviously, the numerical solutions are nearly identical to the analytical ones derived via Eq. (4.41), verifying the validity of the approximate analytical solution. Thus, in view of saving fuels, it is recommended that these critical inclinations should be avoided in Lorentz-augmented hovering practice.

4.2.3.3 Inclined Circular Orbit with Tilted Dipole

The dipole tilt angle of the Earth's magnetic field is about 11.3°. To analyze the effect of the dipole tilt angle on the velocity increment consumption for hovering, a typical scenario is simulated to compare the velocity increment consumption in tilted and nontilted dipole model. The initial phase angle of the magnetic dipole is set as $\Omega_0 = 40°$, and the hovering configuration is set as $\mathbf{\Gamma} = [5\ \text{km}, \pi/4, -\pi/6]$.

Except for the orbital inclination ranging from 0° to 180°, other simulation parameters remain the same as those listed in Table 4.2.

Figure 4.14 depicts the velocity increment consumption for one-orbit hovering of both models. As can be seen, for orbits near the equator, distinctive differences can be observed between the tilted and nontilted dipole model, and the difference decreases as the reference orbit approaches the polar orbit. In view of the tilted dipole model in Eq. (4.8) and the nontilted one in Eq. (4.28), the differences between these two models are larger for orbits near the equator. This explains why the resulting velocity increment consumptions are more different for orbits near the equator.

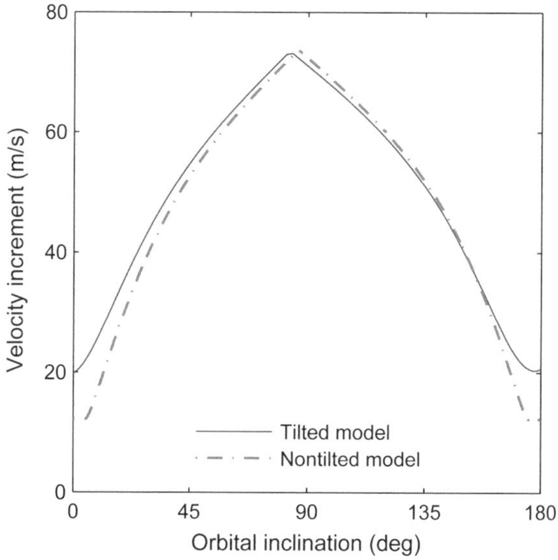

Fig. 4.14 Required velocity increments for the tilted and nontilted dipole model. Reprinted from Ref. [1], Copyright 2013, with permission from Elsevier

4.3 Case of Two-Body Elliptic Reference Orbit

4.3.1 Open-Loop Control

In this section, the target is assumed to be flying in an elliptic orbit. For a Keplerian elliptic orbit, the dynamic equations of the orbital motion are [5]

$$\begin{aligned} \ddot{R}_T - R_T\dot{u}_T^2 &= -\mu/R_T^2 \\ R_T\ddot{u}_T + 2\dot{R}_T\dot{u}_T &= 0 \end{aligned} \tag{4.56}$$

with

$$\begin{aligned} \dot{u}_T &= h_T/R_T^2 \\ \ddot{u}_T &= -2\mu e_T \sin f_T/R_T^3 \\ \dot{R}_T &= \mu e_T \sin f_T/h_T \end{aligned} \tag{4.57}$$

where e_T, f_T, and h_T refer to the eccentricity, true anomaly, and orbital momentum of the target spacecraft, respectively.

In view of Eq. (4.15), the total acceleration for hovering around an elliptic orbit is derived as

$$\boldsymbol{h} = \begin{bmatrix} n_L^2(R_T + x) - n_T^2R_T - \dot{u}_T^2x - \ddot{u}_Ty \\ (n_L^2 - \dot{u}_T^2)y + \ddot{u}_Tx \\ n_L^2z \end{bmatrix} \tag{4.58}$$

Thus, the thruster-generated control acceleration necessary for hovering is

$$\boldsymbol{a}_C = \boldsymbol{h} - \boldsymbol{a}_L = \boldsymbol{h} - \lambda\boldsymbol{l} \tag{4.59}$$

Notably, in the last section dealing with Lorentz-augmented hovering about circular reference orbits, a close range assumption is proposed to derive the approximate analytical results. That is, for spacecraft in close proximity, the magnetic field local at the Lorentz spacecraft can be approximated by that local at the target. However, to derive more precise results, this assumption is omitted in this section, and the real magnetic field local at the Lorentz spacecraft is used.

Suppose that the target's orbit is an inclined elliptic one, and the Earth's magnetic dipole tilt angle is α. Then, the magnetic field local at the Lorentz spacecraft is shown as Eq. (4.6), and the velocity of the Lorentz spacecraft with respect to the local magnetic field is given in Eq. (4.5). Thus, the resulting Lorentz acceleration can be expressed as Eq. (4.11).

Similar to the case with a circular reference orbit, to minimize the consumption of thruster-generated control energy for hovering, a fuel-optimal objective function is chosen as

$$J_a = \Delta V = \int_0^T L[t, \lambda(t)]\mathrm{d}t = \int_0^T ||\boldsymbol{a}_C||\mathrm{d}t \tag{4.60}$$

Then, using similar methods in Sect. 4.2.1.2, the fuel-optimal trajectory of the specific charge of the Lorenz spacecraft can be derived as

$$\lambda^*(f_T) = \begin{cases} (\boldsymbol{h} \cdot \boldsymbol{l})/||\boldsymbol{l}||^2, & ||\boldsymbol{l}|| \neq 0 \\ 0, & ||\boldsymbol{l}|| = 0 \end{cases} \tag{4.61}$$

Also, the optimal trajectory of the thruster-generated control acceleration is

$$\boldsymbol{a}_C^*(f_T) = \begin{cases} \boldsymbol{h} - [(\boldsymbol{h} \cdot \boldsymbol{l})/||\boldsymbol{l}||^2]\boldsymbol{l}, & ||\boldsymbol{l}|| \neq 0 \\ \boldsymbol{h}, & ||\boldsymbol{l}|| = 0 \end{cases} \tag{4.62}$$

Notably, despite that the optimal solutions are similar to those for the case with circular reference orbits, the detailed expressions are not exactly the same. In fact, those of the case with elliptic reference orbits are more complicated. For example, the relative velocity $\boldsymbol{V}_{\mathrm{rel}}$ is also a function of the true anomaly f_T, but the local magnetic field $\boldsymbol{B}$ is a function of both the time t and the true anomaly f_T. Thus, it remains to solve the Keplerian equation to derive the relationship between t and f_T.

4.3.2 Closed-Loop Control

Define $\boldsymbol{X} = [\boldsymbol{\rho}^{\mathrm{T}} \quad \boldsymbol{v}^{\mathrm{T}}]^{\mathrm{T}}$ as the relative state vector, with $\boldsymbol{\rho} = [x \quad y \quad z]^{\mathrm{T}}$ and $\boldsymbol{v} = [\dot{x} \quad \dot{y} \quad \dot{z}]^{\mathrm{T}}$ being the relative position and velocity vector, respectively. Based on the linearized dynamical model of Lorentz-augmented relative motion in Sect. 2.2.1.1, if the hovering distance is negligibly small as compared to the orbital radius of either spacecraft, the relative dynamics between spacecraft can be written in a state-space model as

$$\dot{\boldsymbol{X}}(t) = \boldsymbol{A}(t)\boldsymbol{X}(t) + \boldsymbol{B}\boldsymbol{U}(t) \tag{4.63}$$

with

$$\boldsymbol{A}(t) = \begin{bmatrix} \boldsymbol{0}_{3\times3} & \boldsymbol{I}_{3\times3} \\ \boldsymbol{A}_{v\rho} & \boldsymbol{A}_{vv} \end{bmatrix}, \quad \boldsymbol{B} = \begin{bmatrix} \boldsymbol{0}_{3\times3} \\ \boldsymbol{I}_{3\times3} \end{bmatrix} \tag{4.64}$$

$$\boldsymbol{U}(t) = [a_x + a_R \quad a_y + a_S \quad a_z + a_W]^{\mathrm{T}} \tag{4.65}$$

$$\boldsymbol{A}_{v\rho} = \begin{bmatrix} \dot{u}_T^2 + 2n_T^2 & \ddot{u}_T & 0 \\ -\ddot{u}_T & \dot{u}_T^2 - n_T^2 & 0 \\ 0 & 0 & -n_T^2 \end{bmatrix}, \quad \boldsymbol{A}_{vv} = \begin{bmatrix} 0 & 2\dot{u}_T & 0 \\ -2\dot{u}_T & 0 & 0 \\ 0 & 0 & 0 \end{bmatrix} \tag{4.66}$$

where the expressions of $\dot{u}_T$ and $\ddot{u}_T$ have been given in Eq. (4.57). $\boldsymbol{U}$ denotes the total acceleration consisting of the Lorentz acceleration $\boldsymbol{a}_L = \begin{bmatrix} a_x & a_y & a_z \end{bmatrix}^{\mathrm{T}}$ and the thruster-generated control acceleration $\boldsymbol{a}_C = \begin{bmatrix} a_R & a_S & a_W \end{bmatrix}^{\mathrm{T}}$. $n_T = \sqrt{\mu/R_T^3}$. Also, $\boldsymbol{0}_{M\times N}$ and $\boldsymbol{I}_{M\times N}$ refer to an $M \times N$ zero and identity matrix, respectively.

Considering the linearization errors and the external disturbances that may disturb the hovering position, the nonlinear perturbed dynamical model of Lorentz-augmented hovering can be represented as

$$\dot{\boldsymbol{X}}(t) = \boldsymbol{A}(t)\boldsymbol{X}(t) + \boldsymbol{B}[\boldsymbol{U}(t) + \boldsymbol{D}(\boldsymbol{X}, t)] \tag{4.67}$$

with

$$\boldsymbol{U}(t) = \lambda \boldsymbol{l}(\boldsymbol{X}, t) + \boldsymbol{a}_C(t) \tag{4.68}$$

$$\boldsymbol{D}(\boldsymbol{X}, t) = \boldsymbol{d}(\boldsymbol{X}, t) + \Delta \boldsymbol{F}(\boldsymbol{X}, t) \tag{4.69}$$

where $\boldsymbol{d} = \begin{bmatrix} d_1 & d_2 & d_3 \end{bmatrix}^{\mathrm{T}}$ denotes the bounded external disturbance vector, and $\Delta \boldsymbol{F} = \begin{bmatrix} \Delta F_1 & \Delta F_2 & \Delta F_3 \end{bmatrix}^{\mathrm{T}}$ is the bounded linearization error vector. The sum vector of this two kinds of perturbation is $\boldsymbol{D}(\boldsymbol{X}, t) = \begin{bmatrix} D_1 & D_2 & D_3 \end{bmatrix}^{\mathrm{T}}$ that satisfies $|D_i| \le D_{\max}$, $(i = 1, 2, 3)$.

Denote the desired hovering position as $\boldsymbol{\rho}_d = \begin{bmatrix} x_d & y_d & z_d \end{bmatrix}^{\mathrm{T}}$. According to the definition of hovering, the hovering position remains constant in the LVLH frame. Thus, the desired relative velocity is $\boldsymbol{v}_d = \begin{bmatrix} 0 & 0 & 0 \end{bmatrix}^{\mathrm{T}}$, and the desired relative state vector is $\boldsymbol{X}_d = \begin{bmatrix} \boldsymbol{\rho}_d^{\mathrm{T}} & \boldsymbol{v}_d^{\mathrm{T}} \end{bmatrix}^{\mathrm{T}}$.

In the absence of linearization errors and external perturbation, the ideal desired dynamics is given by

$$\dot{\boldsymbol{X}}_d(t) = \boldsymbol{A}(t)\boldsymbol{X}_d(t) + \boldsymbol{B}\boldsymbol{U}_d(t) \tag{4.70}$$

To achieve hovering, it requires that $\dot{\boldsymbol{X}}_d(t) = \boldsymbol{0}$. Then, in view of $\boldsymbol{B}^{\mathrm{T}}\boldsymbol{B} = \boldsymbol{I}_{3\times 3}$, the desired control input can be obtained as

$$\boldsymbol{U}_d(t) = -\boldsymbol{B}^{\mathrm{T}}\boldsymbol{A}(t)\boldsymbol{X}_d(t) \tag{4.71}$$

Define $\boldsymbol{e}(t) = \boldsymbol{X}(t) - \boldsymbol{X}_d(t) = \begin{bmatrix} \boldsymbol{e}_\rho^{\mathrm{T}} & \boldsymbol{e}_v^{\mathrm{T}} \end{bmatrix}^{\mathrm{T}}$ as the error relative state vector, with $\boldsymbol{e}_\rho = \boldsymbol{\rho} - \boldsymbol{\rho}_d$ and $\boldsymbol{e}_v = \boldsymbol{v} - \boldsymbol{v}_d$ being the error relative position and velocity vector, respectively. Meanwhile, define $\boldsymbol{u}(t) = \boldsymbol{U}(t) - \boldsymbol{U}_d(t)$ as the error control input. Then, by substituting Eq. (4.70) into Eq. (4.67), the error dynamical system can be derived as

$$\dot{\boldsymbol{e}}(t) = \boldsymbol{A}(t)\boldsymbol{e}(t) + \boldsymbol{B}[\boldsymbol{u}(t) + \boldsymbol{D}(\boldsymbol{X}, t)] \tag{4.72}$$

The control objective can now be summarized as eliminating the initial offsets with respect to the desired hovering position, and thereafter maintaining the desired hovering position in the presence of bounded external disturbance and linearization errors.

First, a linear quadratic regulator is designed to minimize the following objective function:

$$J_b = \frac{1}{2}\int_{t_0}^{\infty} \left[\boldsymbol{e}^{\mathrm{T}}(t)\boldsymbol{Q}(t)\boldsymbol{e}(t) + \boldsymbol{u}^{\mathrm{T}}(t)\boldsymbol{R}(t)\boldsymbol{u}(t)\right]\mathrm{d}t \tag{4.73}$$

where $\boldsymbol{Q}(t) = \boldsymbol{Q}^{\mathrm{T}}(t) \in \mathbb{R}^{6\times 6}$ is a semi-positive definite matrix, $\boldsymbol{R}(t) = \boldsymbol{R}^{\mathrm{T}}(t) \in \mathbb{R}^{3\times 3}$ is a positive definite matrix.

For system in Eq. (4.72) without disturbances (i.e., $\boldsymbol{D} = \boldsymbol{0}$), the optimal state feedback control law that minimizes the objective function in Eq. (4.73) is [6]

$$\boldsymbol{u}^*(t) = -\boldsymbol{K}^*(t)\boldsymbol{e}(t) \tag{4.74}$$

with $\boldsymbol{K}^*(t) \in \mathbb{R}^{3\times 6}$ being the optimal feedback matrix, given by

$$\boldsymbol{K}^*(t) = \boldsymbol{R}^{-1}(t)\boldsymbol{B}^{\mathrm{T}}\boldsymbol{P}(t) \tag{4.75}$$

where $\boldsymbol{P}(t) \in \mathbb{R}^{6\times 6}$ is the solution to the following Riccati equation, given by

$$\boldsymbol{P}(t)\boldsymbol{A}(t) + \boldsymbol{A}^{\mathrm{T}}(t)\boldsymbol{P}(t) + \boldsymbol{Q}(t) - \boldsymbol{P}(t)\boldsymbol{B}\boldsymbol{R}^{-1}(t)\boldsymbol{B}^{\mathrm{T}}\boldsymbol{P}(t) = -\dot{\boldsymbol{P}}(t) \tag{4.76}$$

Substitution of the optimal control law into unperturbed system yields that

$$\dot{\boldsymbol{e}}(t) = [\boldsymbol{A}(t) - \boldsymbol{B}\boldsymbol{K}^*(t)]\boldsymbol{e}(t) \tag{4.77}$$

The solution to Eq. (4.77) is then defined as the optimal system trajectory.

Then, an integral sliding mode controller is proposed to ensure the optimal trajectory tracking in the presence of bounded disturbance $\boldsymbol{D}$. An integral sliding surface is designed as [7]

$$\boldsymbol{s}(\boldsymbol{e}, t) = \boldsymbol{C}\boldsymbol{e}(t) + \xi(t) \tag{4.78}$$

$$\dot{\xi}(t) = -\boldsymbol{C}[\boldsymbol{A}(t)\boldsymbol{e}(t) + \boldsymbol{B}\boldsymbol{u}^*(t)], \quad \xi(0) = -\boldsymbol{C}\boldsymbol{e}(0) \tag{4.79}$$

where $\boldsymbol{C} \in \mathbb{R}^{3\times 6}$ is a parameter matrix. Obviously, it holds that $\boldsymbol{s}(\boldsymbol{e}, 0) = \boldsymbol{0}$, indicating that the sliding mode motion starts from the initial time instant.

Taking the time derivative of the sliding surface yields that

$$\dot{s}(\boldsymbol{e},t) = \boldsymbol{C}\dot{\boldsymbol{e}}(t) + \dot{\xi}(t) = \boldsymbol{CB}[\boldsymbol{u}(t) + \boldsymbol{D} - \boldsymbol{u}^*(t)] \tag{4.80}$$

By evaluation of $\dot{s} = \mathbf{0}$, the equivalent control $\boldsymbol{u}_{\text{eq}}$ can be derived as

$$\boldsymbol{u}_{\text{eq}} = \boldsymbol{u}^* = -\boldsymbol{K}^*(t)\boldsymbol{e}(t) \tag{4.81}$$

As can be seen, the equivalent control is identical to the optimal control, thus ensuring the optimal trajectory tracking.

Then the switch control or reaching law $\boldsymbol{u}_s$ is designed as

$$\dot{\boldsymbol{s}} = \boldsymbol{u}_s = -\boldsymbol{\varepsilon}\,\text{sgn}(\boldsymbol{s}) \tag{4.82}$$

where $\boldsymbol{\varepsilon} = \text{diag}(\varepsilon_1, \varepsilon_2, \varepsilon_3) \in \mathbb{R}^{3\times 3}$ is a diagonal parameter matrix, with the coefficient ε_i satisfying $\varepsilon_i > D_i$, $(i = 1, 2, 3)$. $\text{sgn}(\mathbf{s}) = [\,\text{sgn}(s_1) \quad \text{sgn}(s_2) \quad \text{sgn}(s_3)\,]^{\text{T}}$ is the sign function vector.

Let $\boldsymbol{C} = \boldsymbol{B}^{\text{T}}$, then it holds that $\boldsymbol{CB} = \boldsymbol{I}_{3\times 3}$. Thus, the error control input can be summarized as

$$\boldsymbol{u}(t) = \boldsymbol{u}_{\text{eq}}(t) + \boldsymbol{u}_s(t) = -\boldsymbol{K}^*(t)\boldsymbol{e}(t) - \boldsymbol{\varepsilon}\,\text{sgn}(\boldsymbol{s}) \tag{4.83}$$

Also, in combination of the desired control input, the closed-loop control law for Lorentz-augmented spacecraft hovering is derived as

$$\boldsymbol{U}(t) = \boldsymbol{U}_d(t) + \boldsymbol{u}(t) = -\boldsymbol{B}^{\text{T}}\boldsymbol{A}(t)\boldsymbol{X}_d(t) - \boldsymbol{K}^*(t)\boldsymbol{e}(t) - \boldsymbol{\varepsilon}\,\text{sgn}(\boldsymbol{s}) \tag{4.84}$$

Note that $\boldsymbol{U}$ is the total control input consisting of the Lorentz acceleration and the thruster-generated control acceleration. However, the actual control inputs are the specific charge λ and thruster-generated control acceleration $\boldsymbol{a}_C$. Thus, given that propellant is a limited resource on board, it remains to distribute these two kinds of control inputs in a fuel-optimal way. The fuel-optimal objective function is also selected as Eq. (4.60). Likewise, the closed-loop optimal trajectory of the specific charge of Lorentz spacecraft is derived as

$$\lambda = \begin{cases} (\boldsymbol{U}\cdot\boldsymbol{l})/||\boldsymbol{l}||^2, & ||\boldsymbol{l}|| \neq 0 \\ 0, & ||\boldsymbol{l}|| = 0 \end{cases} \tag{4.85}$$

Also, the closed-loop optimal thruster-generated control acceleration is

$$\boldsymbol{a}_C = \begin{cases} \boldsymbol{U} - [(\boldsymbol{U}\cdot\boldsymbol{l})/||\boldsymbol{l}||^2]\boldsymbol{l}, & ||\boldsymbol{l}|| \neq 0 \\ \boldsymbol{U}, & ||\boldsymbol{l}|| = 0 \end{cases} \tag{4.86}$$

Furthermore, it is notable that the near-term feasible maximal specific charge is about 0.03 C/kg. Thus, when the required specific charge exceeds the maximum, that is, $|\lambda| > \lambda_{\max} = 0.03$, the control law is revised as

$$\lambda = \lambda_{\max}\,\mathrm{sgn}(\lambda), \quad \boldsymbol{a}_C = \boldsymbol{U} - \lambda \boldsymbol{l} \tag{4.87}$$

The stability analysis of the closed-loop system is now summarized in the following theorem.

Theorem 4.2 *For the dynamical system of Lorentz-augmented spacecraft hovering in* Eq. (4.67), *if the control laws are designed as* Eqs. (4.85) and (4.86), *then the closed-loop system is asymptotically stable.*

Proof Consider the Lyapunov candidate $V = \boldsymbol{s}^{\mathrm{T}}\boldsymbol{s}/2$, $\forall \boldsymbol{s} \neq \boldsymbol{0}$. Taking the time derivative of V along the system trajectory yields that

$$\begin{aligned} \dot{V} &= \boldsymbol{s}^{\mathrm{T}}\dot{\boldsymbol{s}} = \boldsymbol{s}^{\mathrm{T}}\left[\boldsymbol{C}(\dot{\boldsymbol{X}} - \dot{\boldsymbol{X}}_d) + \dot{\boldsymbol{\xi}}\right] \\ &= \boldsymbol{s}^{\mathrm{T}}\{\boldsymbol{C}[\boldsymbol{A}\boldsymbol{X} + \boldsymbol{B}(\lambda\boldsymbol{l} + \boldsymbol{a}_C + \boldsymbol{D})] - \boldsymbol{C}(\boldsymbol{A}\boldsymbol{X}_d + \boldsymbol{B}\boldsymbol{U}_d) - \boldsymbol{C}(\boldsymbol{A}\boldsymbol{e} + \boldsymbol{B}\boldsymbol{u}^*)\} \end{aligned} \tag{4.88}$$

Substitution of the control laws Eqs. (4.85) and (4.86) into Eq. (4.88) yields

$$\dot{V} = \boldsymbol{s}^{\mathrm{T}}[\boldsymbol{D} - \boldsymbol{\varepsilon}\,\mathrm{sgn}(\boldsymbol{s})] \leq -\sum_{i=1}^{3} |s_i|(\varepsilon_i - D_i) < 0 \tag{4.89}$$

Thus, $\boldsymbol{s}$ will converge to zero asymptotically, so will the error state $\boldsymbol{e}$. This completes the proof.

Remark 4.1 To eliminate the chattering in sliding mode controller, a common practice is to replace the sign function with the saturation function, given by

$$\mathrm{sat}(s_i, \delta) = \begin{cases} s_i/\delta, & |s_i| < \delta \\ \mathrm{sgn}(s_i), & |s_i| \geq \delta \end{cases}, \quad i = 1, 2, 3 \tag{4.90}$$

where δ is the width of the boundary layer.

Remark 4.2 Due to the replacement with the saturation function, the ideal relay characteristic is kept outside the boundary layer. However, within the boundary layer, it is a high-gain feedback control. Thus, the system trajectory will remain in the boundary layer, but not converge exactly to zero as for the case with sign function [8, 9]. Nevertheless, if the boundary layer is thin enough, then the state errors can be kept in acceptable tolerances. But, a too thin boundary layer will incur the chattering problem again. Thus, when designing of the width of the boundary layer, a balance should be kept between the control accuracy and the chattering elimination.

Remark 4.3 This controller, which is designed for Lorentz-augmented spacecraft hovering around elliptic reference orbits, can also be applied to hovering around circular reference orbits, as stated in Sect. 4.2.2.

4.3.3 *Numerical Simulation*

4.3.3.1 Open-Loop Control

It is assumed that the target is flying in an elliptic LEO with a perigee altitude of 500 km and an apogee altitude of 2000 km. The initial orbit elements of the target are listed in Table 4.3. The initial phase angle of the magnetic dipole is set as $\Omega_0 = -60°$, and the desired hovering position is set as $\boldsymbol{\rho}_d = [1.0 \quad -0.5 \quad 0.3]^{\mathrm{T}}$ km.

For a hovering Lorentz spacecraft, the fuel-optimal trajectory of the specific charge during the first orbit is depicted in Fig. 4.15. As can be seen, the specific charge is generally on the order of 10^{-2} C/kg, smaller than the near-term feasible maximum of 0.03 C/kg.

Figure 4.16 compares the thruster-generated control acceleration necessary for hovering of both the charged and uncharged spacecraft. Obviously, with Lorentz force as auxiliary propulsion, the control acceleration can be effectively reduced. For a traditional uncharged spacecraft, the total acceleration necessary for hovering is fully provided by the thrusters on board, that is, $\boldsymbol{a}_C = \boldsymbol{h}$. Notably, for hovering about circular orbits at the desired relative position $\boldsymbol{\rho}_d$, the total acceleration $\boldsymbol{h}$ is constant. However, for hovering about elliptic orbits, $\boldsymbol{h}$ is time-varying, as verified in Fig. 4.16. Furthermore, the velocity increment consumption for one-orbit hovering is about 18.3 m/s for an uncharged spacecraft, while it decreases to 3.25 m/s for a charged Lorentz spacecraft. As can be seen, nearly 82.2 % of the velocity increment can be saved, verifying the advantage of Lorentz spacecraft in saving fuels.

To analyze the effect of the reference orbit eccentricity on Lorentz-augmented hovering, two variables are defined. That is, the mean and maximal specific charge necessary for hovering during the first orbit, denoted by $|\lambda|_{\mathrm{mean}}$ and $|\lambda|_{\max}$, respectively.

The hovering position is also set as $\boldsymbol{\rho}_d = [1.0 \quad -0.5 \quad 0.3]^{\mathrm{T}}$ km. The perigee altitude is fixed at 500 km, but the apogee altitude increases from 500 km. Thus, the corresponding eccentricity increases from zero. Figure 4.17 shows the mean and maximal specific charges necessary for a hovering Lorentz spacecraft at different reference orbit eccentricities. The corresponding consumptions of velocity

Table 4.3 Initial orbit elements of the target spacecraft

Orbit element	Value
Semi-major axis (km)	7628.137
Eccentricity	0.098
Inclination (deg)	15
Right ascension of ascending node (deg)	30
Argument of perigee (deg)	0
True anomaly (deg)	0

Reprinted from Ref. [2], Copyright 2014, with permission from Elsevier

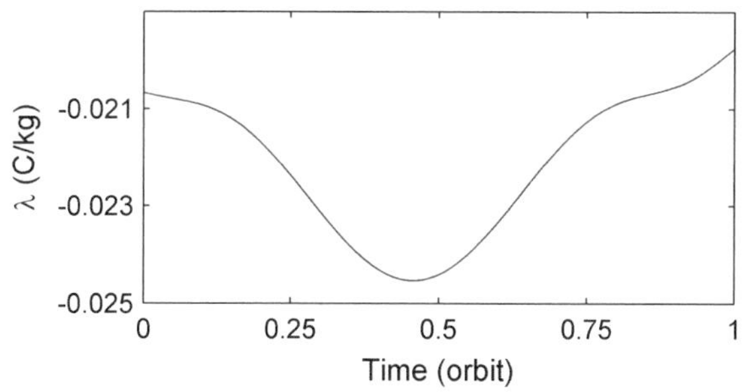

Fig. 4.15 Fuel-optimal trajectory of the specific charge of a hovering Lorentz spacecraft. Reprinted from Ref. [2], Copyright 2014, with permission from Elsevier

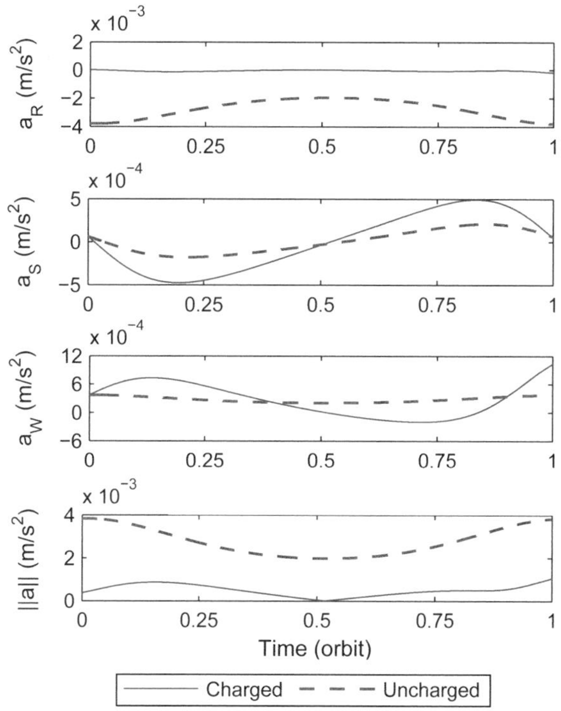

Fig. 4.16 Comparisons of the thruster-generated control accelerations for the charged and uncharged hovering spacecraft. Reprinted from Ref. [2], Copyright 2014, with permission from Elsevier

increments during the first orbit are shown in Fig. 4.18. As can be seen, the charging level increases with increasing eccentricity. This is because that with increasing eccentricity, the orbital altitudes get higher where the magnetic fields is less intenser and the spacecraft travels slower than LEOs. These factors make it less efficient to generate Lorentz acceleration. Thus, higher charging levels are required for larger orbit eccentricities. Furthermore, as shown in Fig. 4.18, the velocity increment consumption for an uncharged spacecraft decreases with increasing eccentricity. Contrarily, for a charged Lorentz spacecraft, a slight increase can be

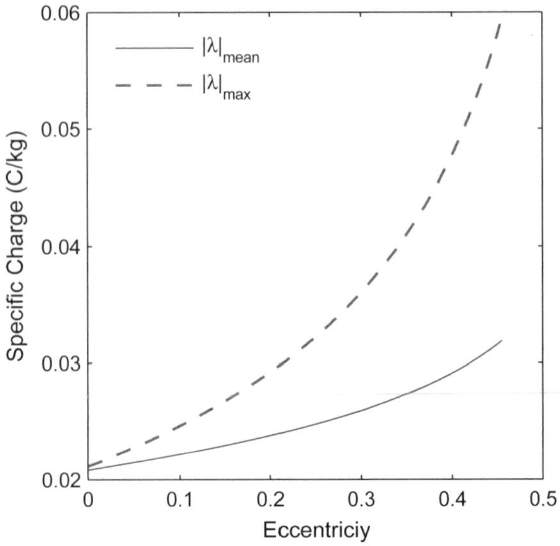

Fig. 4.17 The mean and maximal required specific charges during the first orbital period under different eccentricities. Reprinted from Ref. [2], Copyright 2014, with permission from Elsevier

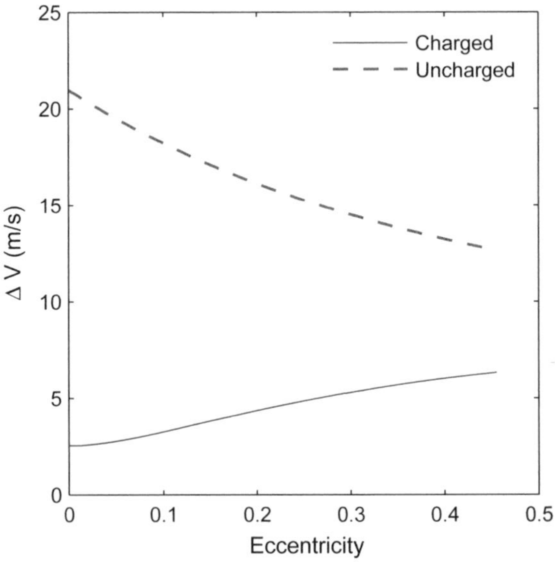

Fig. 4.18 Comparisons of the required velocity increments for the charged and uncharged hovering spacecraft during the first orbital period under different eccentricities. Reprinted from Ref. [2], Copyright 2014, with permission from Elsevier

observed. In spite of the increase, the velocity increment consumption of a charged Lorentz spacecraft is still much smaller than that of an uncharged spacecraft. For example, at the eccentricity of 0.213, nearly 71 % of the velocity increment can be saved.

4.3.3.2 Closed-Loop Control

To verify the feasibility and validity of the closed-loop controller, a typical scenario is simulated in this section. The initial orbit elements of the target are the same as those given in Table 4.3, and the initial phase angle of the magnetic dipole is also chosen as $\Omega_0 = -60°$. The desired hovering position is designated as $\boldsymbol{\rho}_d = [1.0 \quad -0.5 \quad 0.3]^{\mathrm{T}}$ km.

The initial relative position and velocity errors are assumed to be $\boldsymbol{e}_\rho(0) = [50 \quad -100 \quad 50]^{\mathrm{T}}$ m and $\boldsymbol{e}_v(0) = [0.2 \quad 0.5 \quad -0.5]^{\mathrm{T}}$ m/s, respectively. Given that J_2 perturbation, arising from the oblateness of nonhomogeneity of Earth, is one of the most dominant perturbations in LEOs, it is thus included as the external disturbances, and a dynamical model of J_2-perturbed Lorentz spacecraft relative motion is introduced here. Notably, besides J_2 perturbation, another periodic disturbance $\boldsymbol{a}_d$ (unit: m/s^2) is also incorporated as external disturbance, given by

$$\boldsymbol{a}_d = 10^{-5} \begin{bmatrix} 0.5 \sin \bar{n}t \\ 0.5 \sin 2\bar{n}t \\ \sin \bar{n}t \end{bmatrix} \tag{4.91}$$

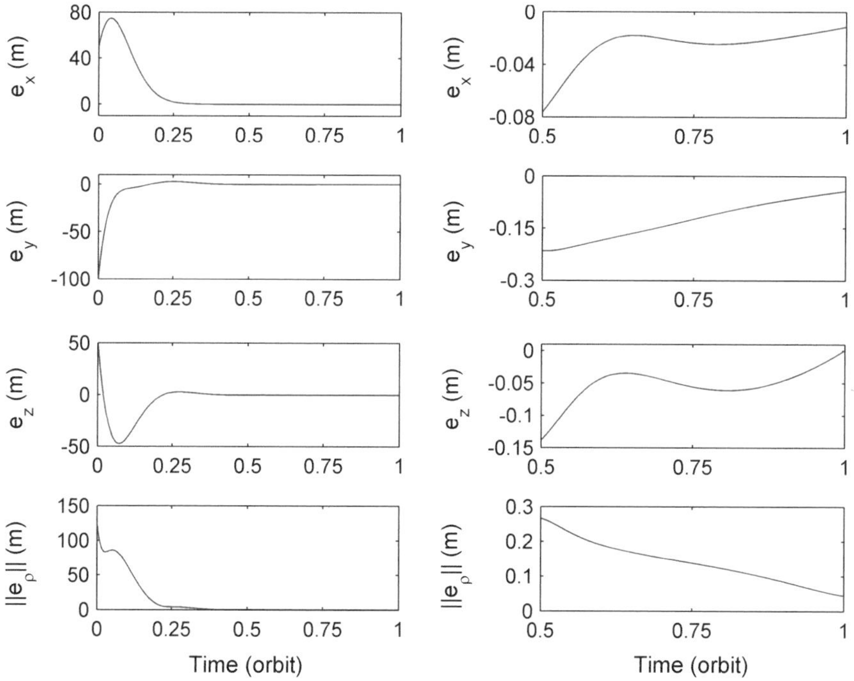

Fig. 4.19 Time histories of relative position errors. Reprinted from Ref. [2], Copyright 2014, with permission from Elsevier

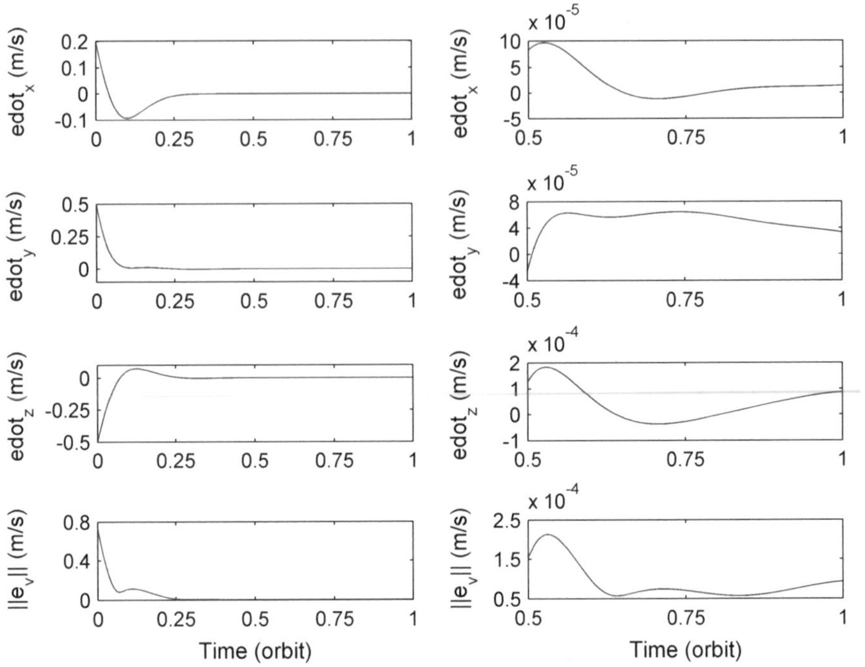

Fig. 4.20 Time histories of relative velocity errors. Reprinted from Ref. [2], Copyright 2014, with permission from Elsevier

where $\bar{n} = \sqrt{\mu / a_T^3}$ is the mean motion of the reference orbit, with a_T being the semi-major axis.

The matrices $\boldsymbol{Q}$ and $\boldsymbol{R}$ determine the performance of the LQR. Given that propellant is a limit resource on board, the control input should be given priority. Thus, the matrices $\boldsymbol{Q}$ and $\boldsymbol{R}$ are designed as

$$\boldsymbol{Q} = \boldsymbol{I}_{6\times 6}, \quad \boldsymbol{R} = \mathrm{diag}(10^{10}, 10^{10}, 10^{10}) \tag{4.92}$$

Furthermore, $\boldsymbol{\varepsilon} = 5 \times 10^{-5} \boldsymbol{I}_{3\times 3}$, and $\delta = 10^{-4}$.

Time histories of the relative position and velocity errors are shown in Figs. 4.19 and 4.20, respectively. As can be seen, both relative position and velocity errors converge to the neighborhood of zero after about 0.4 T, where T refers to the orbital period. Details during the last half period are also enlarged in the right sides of both figures to show the steady errors more distinctively. The terminal control accuracy of the relative position is on the order of 10^{-2} m, and that of the relative velocity is on the order of 10^{-4} m/s. Simulation results verify the feasibility of the closed-loop control scheme.

Figure 4.21 depicts the time histories of the total acceleration necessary for hovering. For an uncharged spacecraft, the total acceleration is fully provided by

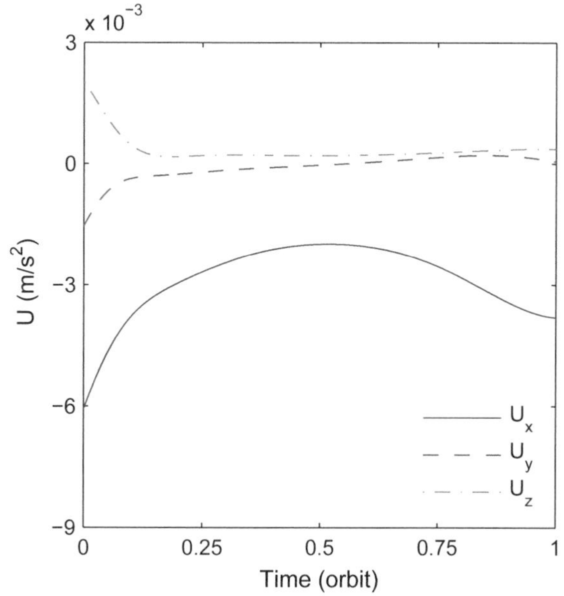

Fig. 4.21 Time histories of total control accelerations. Reprinted from Ref. [2], Copyright 2014, with permission from Elsevier

thrusters on board, that is, $\boldsymbol{a}_C = \boldsymbol{U}$. Thus, the velocity increment consumption for an uncharged spacecraft is 19.2 m/s. Compared to the open-loop case in Sect. 4.3.3.1, a bit more velocity increment is consumed to eliminate the initial relative state errors and compensate for the external disturbances and linearization errors. Differently, by using a charged Lorentz spacecraft, the velocity increment consumption could be effectively reduced. Using the fuel-optimal distribution laws in Eqs. (4.85) and (4.86), the distributed control inputs of a Lorentz spacecraft are shown in Fig. 4.22. At the initial beginnings (about 0.02 T, 115 s), the required specific charge exceeds the maximum of 0.03 C/kg, it is thus set as the maximum with corresponding polarity, as shown in the right side of Fig. 4.22. As can be seen, after about 0.4 T, the real control inputs begin to track the nominal desired ones. Also, the Lorentz spacecraft consumes a velocity increment of nearly 3.98 m/s to complete the closed-loop hovering control. Thus, nearly 79.3 % of the velocity increment can be saved as compared to an uncharged spacecraft, proving the advantage of the Lorentz spacecraft in saving fuels again.

Furthermore, time histories of the relative J_2-perturbation acceleration between spacecraft are shown in Fig. 4.23. As can be seen, the relative J_2-perturbation acceleration is generally on the order of 10^{-6} m/s^2.

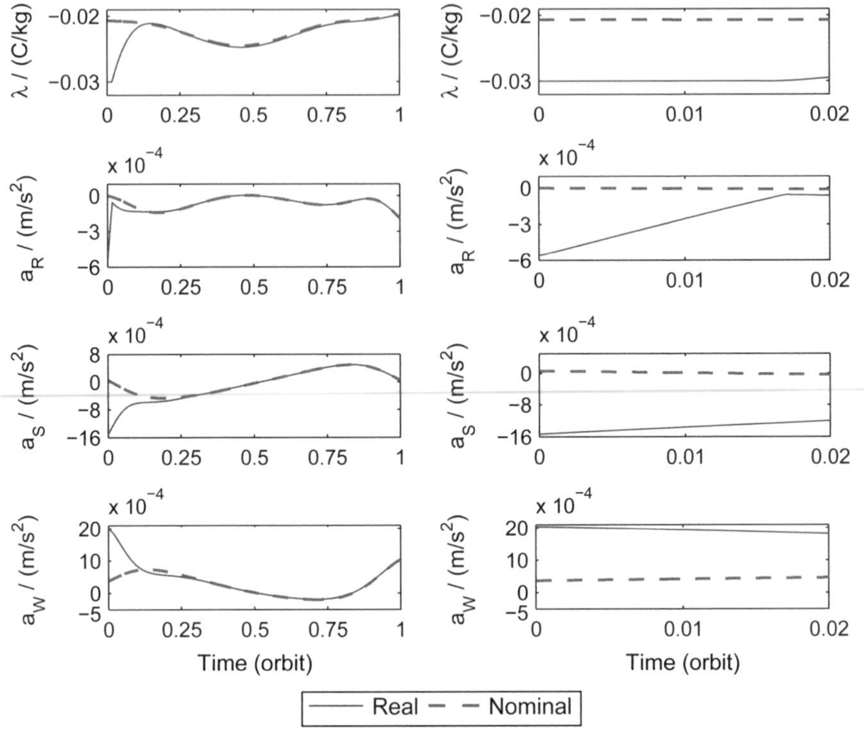

Fig. 4.22 Time histories of real and nominal control inputs. Reprinted from Ref. [2], Copyright 2014, with permission from Elsevier

4.4 Case of J_2-Perturbed Reference Orbit

4.4.1 Dynamical Model

According to the analyses in Sect. 2.2.3.1, the equations of J_2-perturbed Lorentz spacecraft relative motion can be summarized as [10]

$$\begin{cases} \ddot{x} = 2\dot{y}\omega_z - x(\eta_L^2 - \omega_z^2) + y\varepsilon_z - z\omega_x\omega_z - (\xi_L - \xi)\sin i_T \sin u_T \\ \qquad -R_T(\eta_L^2 - \eta^2) + a_x + a_R \\ \ddot{y} = -2\dot{x}\omega_z + 2\dot{z}\omega_x - x\varepsilon_z - y(\eta_L^2 - \omega_z^2 - \omega_x^2) + z\varepsilon_x \\ \qquad -(\xi_L - \xi)\sin i_T \cos u_T + a_y + a_S \\ \ddot{z} = -2\dot{y}\omega_x - x\omega_x\omega_z - y\varepsilon_x - z(\eta_L^2 - \omega_x^2) - (\xi_L - \xi)\cos i_T \\ \qquad + a_z + a_W \end{cases} \tag{4.93}$$

with [11]

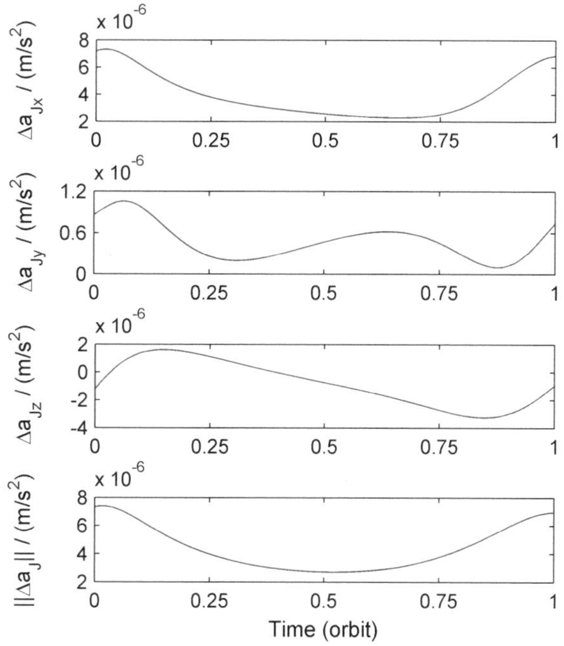

Fig. 4.23 Time histories of relative J_2-perturbation acceleration. Reprinted from Ref. [2], Copyright 2014, with permission from Elsevier

$$\eta_L^2 = \frac{\mu}{R_L^3} + \frac{k_J}{R_L^5} - \frac{5k_J R_{Lz}^2}{R_L^7}, \quad \eta^2 = \frac{\mu}{R_T^3} + \frac{k_J}{R_T^5} - \frac{5k_J \sin^2 i_T \sin^2 u_T}{R_T^5} \tag{4.94}$$

$$\xi_L = \frac{2k_J R_{Lz}}{R_L^5}, \quad \xi = \frac{2k_J \sin i_T \sin u_T}{R_T^4} \tag{4.95}$$

$\boldsymbol{\omega} = [\omega_x \quad \omega_y \quad \omega_z]^{\mathrm{T}}$ and $\boldsymbol{\varepsilon} = [\varepsilon_x \quad \varepsilon_y \quad \varepsilon_z]^{\mathrm{T}}$ are, respectively, the orbital angular velocity and acceleration of the LVLH frame, given by [11]

$$\begin{bmatrix} \omega_x \\ \omega_y \\ \omega_z \end{bmatrix} = \begin{bmatrix} -(k_J \sin 2i_T \sin u_T)/(h_T R_T^3) \\ 0 \\ h_T/R_T^2 \end{bmatrix} \tag{4.96}$$

$$\begin{bmatrix} \varepsilon_x \\ \varepsilon_y \\ \varepsilon_z \end{bmatrix} = \begin{bmatrix} \frac{3V_r k_J \sin 2i_T \sin u_T}{R_T^4 h_T} - \frac{k_J \sin 2i_T \cos u_T}{R_T^5} - \frac{8k_J^2 \sin^3 i_T \cos i_T \sin^2 u_T \cos u_T}{R_T^6 h_T^2} \\ 0 \\ -\frac{2h_T V_r}{R_T^3} - \frac{k_J \sin^2 i_T \sin 2u_T}{R_T^5} \end{bmatrix} \tag{4.97}$$

where $R_{Lz} = (R_T + x)\sin i_T \sin u_T + y \sin i_T \cos u_T + z \cos i_T$ is the projection of $\boldsymbol{R}_L$ on the Z_I axis of ECI frame. The coefficient k_J is defined as $k_J = 3J_2\mu R_E^2/2$, with R_E being the radius of Earth.

$\boldsymbol{a}_C = [a_R \quad a_S \quad a_W]^{\mathrm{T}}$ and $\boldsymbol{a}_L = \lambda \boldsymbol{l} = \lambda [l_x \quad l_y \quad l_z]^{\mathrm{T}}$ are the thruster-generated control acceleration and Lorentz acceleration acting on the Lorentz spacecraft, respectively. The expression of $\boldsymbol{l}$ is given by

$$\begin{aligned} l_x = & (B_0/R_L^5)V_{y,J_2}\{(3z^2 - R_L^2)n_z + 3z[(R_T + x)n_x + yn_y]\} \\ & - (B_0/R_L^5)V_{z,J_2}\{(3y^2 - R_L^2)n_y + 3y[(R_T + x)n_x + zn_z]\} \end{aligned} \tag{4.98}$$

$$\begin{aligned} l_y = & (B_0/R_L^5)V_{z,J_2}\left\{[3(R_T + x)^2 - R_L^2]n_z + 3(R_T + x)(yn_y + zn_z)\right\} \\ & - (B_0/R_L^5)V_{x,J_2}\{(3z^2 - R_L^2)n_z + 3z[(R_T + x)n_x + zn_z]\} \end{aligned} \tag{4.99}$$

$$\begin{aligned} l_z = & (B_0/R_L^5)V_{x,J_2}\{(3y^2 - R_L^2)n_y + 3y[(R_T + x)n_x + zn_z]\} \\ & - (B_0/R_L^5)V_{y,J_2}\left\{[3(R_T + x)^2 - R_L^2]n_z + 3(R_T + x)(yn_y + zn_z)\right\} \end{aligned} \tag{4.100}$$

with

$$\begin{bmatrix} n_x \\ n_y \\ n_z \end{bmatrix} = \begin{bmatrix} -(\cos\beta\cos u_T + \sin\beta\cos i_T\sin u_T)\sin\alpha - \sin i_T\sin u_T\cos\alpha \\ (\cos\beta\sin u_T - \sin\beta\cos i_T\cos u_T)\sin\alpha - \sin i_T\cos u_T\cos\alpha \\ \sin\beta\sin i_T\sin\alpha - \cos i_T\cos\alpha \end{bmatrix} \tag{4.101}$$

$$\begin{bmatrix} V_{x,J_2} \\ V_{y,J_2} \\ V_{z,J_2} \end{bmatrix} = \begin{bmatrix} \dot{R}_T + \dot{x} - y(\omega_z - \omega_E\cos i_T) - z\omega_E\sin i_T\cos u_T \\ \dot{y} + (R_T + x)(\omega_z - \omega_E\cos i_T) - z(\omega_x - \omega_E\sin i_T\sin u_T) \\ \dot{z} + (R_T + x)\omega_E\sin i_T\cos u_T + y(\omega_x - \omega_E\sin i_T\sin u_T) \end{bmatrix} \tag{4.102}$$

where $\beta = \Omega_M - \Omega_T$ and $\Omega_M = \omega_E t + \Omega_0$. R_T, V_r, h_T, i_T, u_T, and Ω_T refer to the orbital radius, radial velocity, orbital angular momentum, inclination, argument of latitude, and right ascension of the ascending node of the target spacecraft, respectively. The dynamics of these orbit parameters is governed by [11]

$$\begin{cases} \dot{R}_T = V_r \\ \dot{V}_r = -\mu/R_T^2 + h_T^2/R_T^3 - k_J(1 - 3\sin^2 i_T\sin^2 u_T)/R_T^4 \\ \dot{h}_T = -(k_J\sin^2 i_T\sin 2u_T)/R_T^3 \\ \dot{i}_T = -(k_J\sin 2i_T\sin 2u_T)/(2h_T R_T^3) \\ \dot{u}_T = h_T/R_T^2 + (2k_J\cos^2 i_T\sin^2 u_T)/(h_T R_T^3) \\ \dot{\Omega}_T = -(2k_J\cos i_T\sin^2 u_T)/(h_T R_T^3) \end{cases} \tag{4.103}$$

4.4.2 Open-Loop Control

According to the definition of hovering, the hovering position remains constant in the LVLH frame. Thus, the time derivative of the relative position vector is zero in the LVLH frame, given by

$$\dot{x} = \dot{y} = \dot{z} = 0, \quad \ddot{x} = \ddot{y} = \ddot{z} = 0 \tag{4.104}$$

Substitution of Eq. (4.104) into Eq. (4.93) yields that

$$\boldsymbol{a}_C = \boldsymbol{h} - \lambda \boldsymbol{l} \tag{4.105}$$

where $\boldsymbol{h}$ is the total acceleration necessary for hovering, given by

$$\boldsymbol{h} = \begin{bmatrix} x(\eta_L^2 - \omega_z^2) - y\varepsilon_z + z\omega_x\omega_z + (\xi_L - \xi)\sin i_T \sin u_T + R_T(\eta_L^2 - \eta^2) \\ x\varepsilon_z + y(\eta_L^2 - \omega_z^2 - \omega_x^2) - z\varepsilon_x + (\xi_L - \xi)\sin i_T \cos u_T \\ x\omega_x\omega_z + y\varepsilon_x + z(\eta_L^2 - \omega_x^2) + (\xi_L - \xi)\cos i_T \end{bmatrix} \tag{4.106}$$

Likewise, using similar methods in Sect. 4.3.1, the fuel-optimal trajectory of the specific charge of Lorentz spacecraft can be derived as

$$\lambda^* = \begin{cases} (\boldsymbol{h} \cdot \boldsymbol{l})/||\boldsymbol{l}||^2, & ||\boldsymbol{l}|| \neq 0 \\ 0, & ||\boldsymbol{l}|| = 0 \end{cases} \tag{4.107}$$

Similarly, the optimal thruster-generated control acceleration is

$$\boldsymbol{a}_C^* = \begin{cases} \boldsymbol{h} - [(\boldsymbol{h} \cdot \boldsymbol{l})/||\boldsymbol{l}||^2]\boldsymbol{l}, & ||\boldsymbol{l}|| \neq 0 \\ \boldsymbol{h}, & ||\boldsymbol{l}|| = 0 \end{cases} \tag{4.108}$$

4.4.3 Numerical Simulation

This section will analyze the effect of J_2 perturbation on Lorentz-augmented spacecraft hovering by comparisons of the numerical results derived in the two-body and J_2-perturbed models.

The target is assumed to be flying in a near-circular orbit. In the absence of perturbations, the ideal orbital period is about 1.6 h. The initial orbital elements of the target are listed in Table 4.4, and the initial phase angle of the magnetic dipole is set as $\Omega_0 = -60°$.

The desired hovering position is set as $\boldsymbol{\rho} = \begin{bmatrix} 1.0 & -1.0 & 0.5 \end{bmatrix}^{\mathrm{T}}$ km. Time histories of the specific charge necessary for a hovering Lorentz spacecraft in the J_2-perturbed environment are shown in Fig. 4.24. As can be seen, the required specific

Table 4.4 Initial orbit elements of the target spacecraft

Orbit element	Value
Semi-major axis (km)	6945.034
Eccentricity	0.0002
Inclination (deg)	30
Right ascension of ascending node (deg)	50
Argument of perigee (deg)	20
True anomaly (deg)	−20

Reprinted from Ref. [10], Copyright 2015, with permission from SAGE

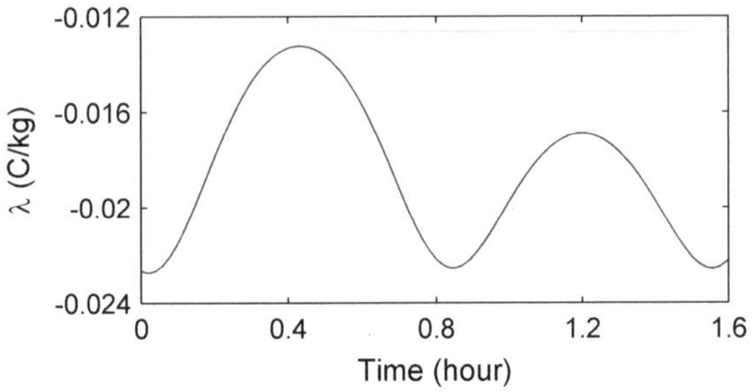

Fig. 4.24 Time histories of the required specific charge of J_2-perturbed Lorentz spacecraft. Reprinted from Ref. [10], Copyright 2015, with permission from SAGE

charge is generally on the order of 10^{-2} C/kg, smaller than the near-term feasible maximum of 0.03 C/kg. For the charged and uncharged spacecraft, the required thruster-generated control acceleration is compared in Fig. 4.25. For an uncharged spacecraft, the total acceleration necessary for hovering is fully provided by the thrusters, that is, $\boldsymbol{a}_C = \boldsymbol{h}$. The velocity increment consumption for an uncharged spacecraft is about 20.95 m/s, while it decreases to 7.95 m/s for a charged Lorentz spacecraft, with nearly 62 % of the velocity increment saved.

Hovering is sensitive to external perturbations. To further analyze the effect of J_2 perturbation on Lorentz-augmented spacecraft hovering, the optimal trajectories of the specific charge and thruster-generated control acceleration derived in an unperturbed environment are introduced for comparisons. The differential control inputs between the J_2-perturbed and unperturbed model are shown in Fig. 4.26. The total simulation time is 24 h, which is about 15 orbital periods. The differential specific charge fluctuates about zero, and the fluctuation amplitude is about one order of magnitude smaller than that of the required specific charge. The maximal specific charge error is about 5.8×10^{-3} C/kg. Likewise, the differential control accelerations are predominantly oscillatory with increasing amplitudes. The maximal radial, in-track, and normal control accelerations are 7.6×10^{-4}, 4.0×10^{-5}, and 1.1×10^{-3} m/s^2, respectively, which are almost on the same orders of magnitude as those of the control acceleration on each axis. Furthermore, all errors remain small during the first orbit (about 1.6 h), and thereafter begin to increase

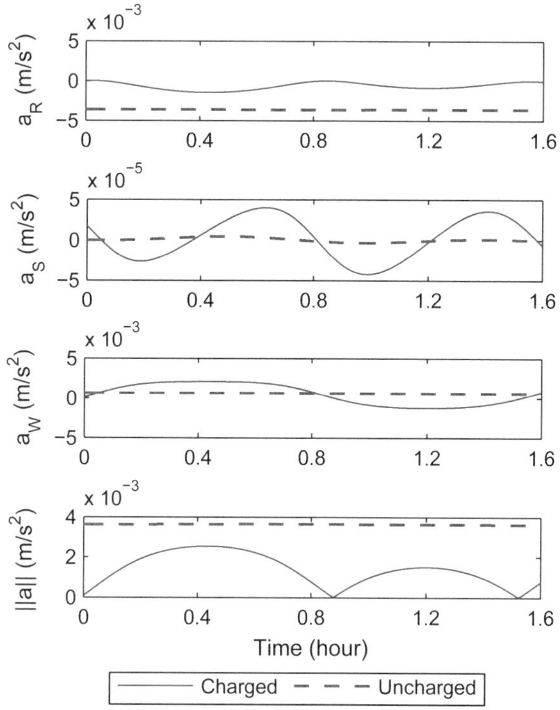

Fig. 4.25 Comparisons on control acceleration of J_2-perturbed charged and uncharged spacecraft. Reprinted from Ref. [10], Copyright 2015, with permission from SAGE

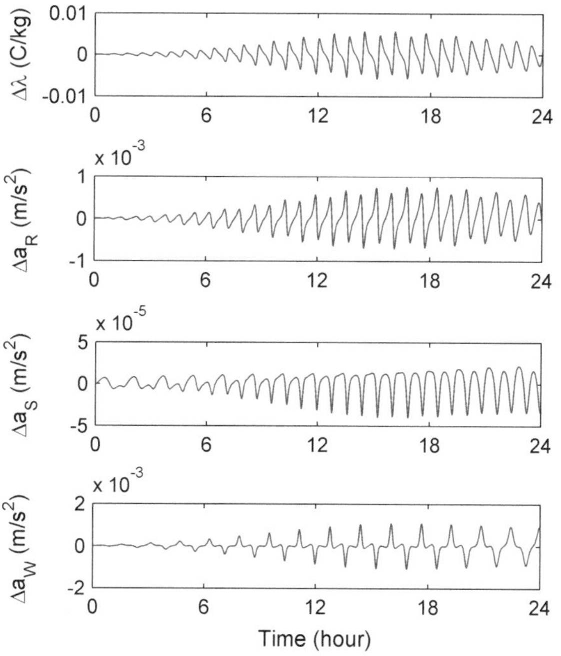

Fig. 4.26 Errors between the control inputs derived from the J_2-perturbed and unperturbed models. Reprinted from Ref. [10], Copyright 2015, with permission from SAGE

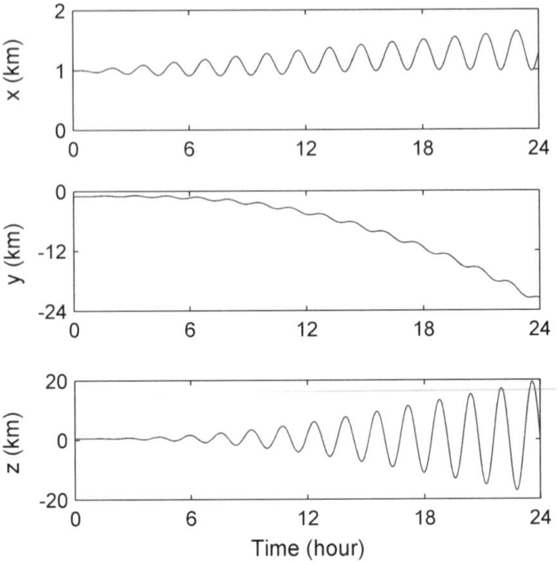

Fig. 4.27 Time histories of J_2-perturbed hovering position with control inputs derived via unperturbed model. Reprinted from Ref. [10], Copyright 2015, with permission from SAGE

significantly. Such distinctive errors will definitely result in the drift of the desired hovering position.

To verify this argument, the required hybrid control inputs derived in the unperturbed model are substituted into the J_2-perturbed model (i.e., Eqs. (4.93)–(4.103)). The trajectory of the hovering position is then derived via numerical integrations of the nonlinear dynamic equations by using forth-order Runge–Kutta method, as shown in Fig. 4.27. The time step used in the Runge–Kutta method is set as 0.01 s. As can be seen, if the hybrid control inputs obtained via an unperturbed model are directly applied in the J_2-perturbed model, the radial hovering position is mainly oscillatory with slight increasing amplitude. The in-track hovering position drifts significantly, and the normal hovering position fluctuates with increasing amplitude. During the one-day hovering, the maximal offsets on each axis are about 0.64, 20.61, and 18.71 km, respectively. Obviously, the in-track hovering position drifts most significantly, followed by the normal one. By contrast, the radial hovering position drifts a bit during one day. Therefore, it can be concluded that when designing controllers for long-term Lorentz-augmented hovering in LEOs, the J_2 perturbation should be taken into account.

4.5 Conclusions

This chapter investigates the feasibility of using the geomagnetic Lorentz as primary propulsion for spacecraft hovering. Both circular and elliptic reference orbits are considered. First, propellantless hovering configurations are derived via an approximate analytic method. For other configurations that necessitate the hybrid inputs consisting of specific charge and thruster-generated control acceleration, optimal open-loop distribution laws of these two kinds of propulsion are derived via the Lagrange method. Then, a closed-loop integral sliding mode controller is proposed to track the optimal trajectory in the presence of initial errors, linearization errors, and external perturbations. Numerical simulations verify the validity of the proposed open-loop and closed-loop control methods. The results indicate that a specific charge on the order of 10^{-1} C/kg could afford a hovering distance of about several kilometers in typical LEOs.

Also, given that J_2 perturbation is one of the most dominant perturbations in LEOs, a dynamical model of J_2-perturbed Lorentz spacecraft relative motion is introduced for comparative study. The comparison results indicate that J_2 perturbation should be taken into account when analyzing long-term Lorentz-augmented hovering in LEOs.

References

1. Huang X, Yan Y, Zhou Y (2014) Dynamics and control of spacecraft hovering using the geomagnetic Lorentz force. Adv Space Res 53:518–531
2. Huang X, Yan Y, Zhou Y et al (2014) Sliding mode control of Lorentz-augmented spacecraft hovering around elliptic orbits. Acta Astronaut 103:257–268
3. Pollock GE, Gangestad JW, Longuski JM (2010) Inclination change in low-Earth orbit via the geomagnetic Lorentz force. J Guid Control Dyn 33:1387–1395
4. Finlay CC, Maus S, Beggan CD et al (2010) International geomagnetic reference field: the eleventh generation. Geophys J Int 183:1216–1230
5. Tsujii S, Bando M, Yamakawa H (2013) Spacecraft formation flying dynamics and control using the geomagnetic Lorentz force. J Guid Control Dyn 36:136–148
6. Zheng D (2002) Linear system theory. Tsinghua, Beijing
7. Imani A, Bahrami M (2013) Optimal sliding mode control for spacecraft formation flying in eccentric orbits. Proc Inst Mech Eng Part I J Syst Control Eng 227:474–481
8. Hung JY, Gao W, Hung JC (1993) Variable structure control: a survey. IEEE Trans Ind Electron 40:2–22
9. Young KD, Utkin VI, Özgüner Ü (1999) A control engineer's guide to sliding mode control. IEEE Trans Control Syst Technol 7:328–342
10. Huang X, Yan Y, Zhou Y et al (2015) Nonlinear relative dynamics of Lorentz spacecraft about J_2-perturbed orbit. Proc Inst Mech Eng Part G J Aerosp Eng 229:467–478. doi:10.1177/0954410014537231
11. Xu G, Wang D (2008) Nonlinear dynamic equations of satellite relative motion around an oblate Earth. J Guid Control Dyn 31:1521–1524

Chapter 5
Dynamics and Control of Lorentz-Augmented Spacecraft Formation Flying

Lorentz-augmented spacecraft formation flying (SFF) is one of the most promising applications of Lorentz spacecraft. As a means of propellantless electromagnetic propulsion, the Lorentz force can be applied to formation establishment, reconfiguration, and maintenance. Based on the dynamical model of the Lorentz-augmented SFF, optimal open-loop control trajectories for formation establishment or reconfiguration are solved by pseudospectral method at first. Both fixed and free terminal conditions are considered to make comparisons. Meanwhile, the effect of the dipole tilt angle on the optimal trajectories is also analyzed numerically. Then, closed-loop controllers are designed to track the optimal trajectory.

5.1 Problem Formulation and Dynamical Model

5.1.1 Problem Formulation

Formation establishment refers to the establishment of a formation configuration from arbitrary initial relative state using orbital maneuvers, and formation reconfiguration refers to the transformation from the existing configuration to another new one using orbital maneuvers. As can be seen, both operations require propulsions to perform orbital maneuvering. Obviously, compared to traditional thrusters that use chemical fuels, the novel propellantless means is more preferable and advantageous. In view of this fact, the feasibility of formation establishment or reconfiguration augmented by the geomagnetic Lorentz force is investigated in this chapter, and problem formulation can be summarized as follows:

1. Design propellantless or nearly propellantless optimal control schemes for formation establishment and reconfiguration via the pseudospectral method. Both fixed and free final conditions are considered to make comparisons. Also, both circular and elliptic reference orbits are considered;

© Springer Science+Business Media Singapore 2017
Y. Yan et al., *Dynamics and Control of Lorentz-Augmented Spacecraft Relative Motion*, DOI 10.1007/978-981-10-2603-4_5

2. Analyze the effect of magnetic dipole tilt angle on the optimal trajectories of formation establishment or reconfiguration propelled by the Lorentz force;
3. Design closed-loop tracking controllers to guarantee optimal trajectory tracking in the presence of initial offsets and external perturbations.

5.1.2 Dynamical Model

The spacecraft involved in a formation are generally referred to as the chief and deputy spacecraft. The deputy is assumed to be a charged Lorentz spacecraft, but the chief is uncharged. The relative dynamics between the chief and deputy is described in the chief's LVLH frame, where the chief's LVLH frame is the same as the target's LVLH frame as defined in Sect. 2.2.1.1. Define the position vector of the deputy relative to the chief as $\boldsymbol{\rho} = \boldsymbol{R}_L - \boldsymbol{R}_C = \begin{bmatrix} x & y & z \end{bmatrix}^{\mathrm{T}}$, where $\boldsymbol{R}_C$ and $\boldsymbol{R}_L$ refer to the orbital radius vector of the chaser and the deputy, respectively. By referring to Sect. 2.2.1.1, the equations of Lorentz-augmented SFF can be rewritten in the LVLH frame as

$$\begin{aligned} \ddot{x} &= 2\dot{u}_C\dot{y} + \dot{u}_C^2 x + \ddot{u}_C y + n_C^2 R_C - n_L^2 (R_C + x) + a_x + a_R \\ \ddot{y} &= -2\dot{u}_C\dot{x} + \dot{u}_C^2 y - \ddot{u}_C x - n_L^2 y + a_y + a_S \\ \ddot{z} &= -n_C^2 z + a_z + a_W \end{aligned} \tag{5.1}$$

where $n_C = \sqrt{\mu/R_C^3}$ and $n_L = \sqrt{\mu/R_L^3}$, R_C and $R_L = [(R_C + x)^2 + y^2 + z^2]^{1/2}$ refer to the orbital radius of the chief and the deputy, respectively. μ is the gravitational parameter of Earth. u_C is the argument of latitude of the chief. Thus, $\dot{u}_C$ and $\ddot{u}_C$ refer to the orbital angular velocity and acceleration of the chief, respectively. $\boldsymbol{a}_C = \begin{bmatrix} a_R & a_S & a_W \end{bmatrix}^{\mathrm{T}}$ is the thruster-generated control acceleration acting on the deputy Lorentz spacecraft, and $\boldsymbol{a}_L = \begin{bmatrix} a_x & a_y & a_z \end{bmatrix}^{\mathrm{T}}$ denotes the Lorentz acceleration, given by

$$\boldsymbol{a}_L = (q_L/m_L)\boldsymbol{V}_{\mathrm{rel}} \times \boldsymbol{B} \tag{5.2}$$

where $\lambda = q_L/m_L$ is the specific charge of Lorentz spacecraft. $\boldsymbol{B}$ is the magnetic field local at the Lorentz spacecraft, given by [1]

$$\boldsymbol{B} = \begin{bmatrix} B_x & B_y & B_z \end{bmatrix}^{\mathrm{T}} = (B_0/R_L^3)[3(\hat{\boldsymbol{n}} \cdot \hat{\boldsymbol{R}}_L)\hat{\boldsymbol{R}}_L - \hat{\boldsymbol{n}}] \tag{5.3}$$

where $\hat{\boldsymbol{R}}_L = (1/R_L)\begin{bmatrix} R_T + x & y & z \end{bmatrix}^{\mathrm{T}}$ is the unit orbital radius vector of the Lorentz spacecraft, and $\hat{\boldsymbol{n}}$ is the unit magnetic dipole momentum vector, given by

$$\hat{\boldsymbol{n}} = \begin{bmatrix} n_x \\ n_y \\ n_z \end{bmatrix} = \begin{bmatrix} -(\cos\beta\cos u_C + \sin\beta\cos i_C\sin u_C)\sin\alpha - \sin i_C\sin u_C\cos\alpha \\ (\cos\beta\sin u_C - \sin\beta\cos i_C\cos u_C)\sin\alpha - \sin i_C\cos u_C\cos\alpha \\ \sin\beta\sin i_C\sin\alpha - \cos i_C\cos\alpha \end{bmatrix} \tag{5.4}$$

where i_C is the orbital inclination of the chief, and α is dipole tilt angle. The angle β is defined as $\beta = \Omega_M - \Omega_C$. Ω_C is the right ascension of the ascending node of the chief, and $\Omega_M = \omega_E t + \Omega_0$ is the inertial phase angle of the magnetic dipole. Also,ω_E is the rotation rate of Earth, t is the time, and Ω_0 is the initial phase angle of the dipole.

Given that the Earth's magnetic field is corotating with Earth, the velocity of the Lorentz spacecraft relative to the magnetic field is thus given by

$$\boldsymbol{V}_{\text{rel}} = \begin{bmatrix} V_x \\ V_y \\ V_z \end{bmatrix} = \begin{bmatrix} \dot{R}_C + \dot{x} - y(\dot{u}_C - \omega_E\cos i_C) - z\omega_E\sin i_C\cos u_C \\ \dot{y} + (R_C + x)(\dot{u}_C - \omega_E\cos i_C) + z\omega_E\sin i_C\sin u_C \\ \dot{z} + (R_C + x)\omega_E\sin i_C\cos u_C - y\omega_E\sin i_C\sin u_C \end{bmatrix} \tag{5.5}$$

Substitution of Eqs. (5.3) and (5.5) into Eq. (5.2) yields the expression of Lorentz acceleration in the LVLH frame. Then, substitution of the expression of $\boldsymbol{a}_L$ into Eq. (5.1) yields the dynamical model of Lorentz-augmented SFF.

5.2 Case of Two-Body Circular Reference Orbit

5.2.1 Formation Configurations in Circular Orbits

A circular reference orbit is considered in this section. As derived in Sect. 2.2.1.1, if the relative distance is negligibly small, the relative dynamics can be linearized as

$$\begin{aligned} \ddot{x} &= 2n_C\dot{y} + 3n_C^2 x + a_x + a_R \\ \ddot{y} &= -2n_C\dot{x} + a_y + a_W \\ \ddot{z} &= -n_C^2 z + a_z + a_S \end{aligned} \tag{5.6}$$

In the absence of control inputs (i.e., $\boldsymbol{a}_L = \boldsymbol{0}$ and $\boldsymbol{a}_C = \boldsymbol{0}$), the analytic solutions to the unforced relative dynamics are [2, 3]

$$x(t) = (\dot{x}_0/n_C)\sin n_C t - (3x_0 + 2\dot{y}_0/n_C)\cos n_C t + 2(2x_0 + \dot{y}_0/n_C) \tag{5.7}$$

$$\begin{aligned} y(t) = {} & (2\dot{x}_0/n_C)\cos n_C t + (6x_0 + 4\dot{y}_0/n_C)\sin n_C t - (6n_C x_0 + 3\dot{y}_0)t \\ & + y_0 - 2\dot{x}_0/n_C \end{aligned} \tag{5.8}$$

$$z(t) = (\dot{z}_0/n_C)\sin n_C t + z_0 \cos n_C t \tag{5.9}$$

where x_0, y_0, and z_0 refer to the initial radial, in-track, and normal relative position, respectively. $\dot{x}_0$, $\dot{y}_0$, and $\dot{z}_0$ refer to the initial radial, in-track, and normal relative velocity, respectively.

To avoid the long-term drift in the in-track direction, the coefficient of the long-term item (i.e., $6n_C x_0 + 3\dot{y}_0$) should be zero. Thus, it holds that

$$\dot{y}_0 = -2n_C x_0 \tag{5.10}$$

Equation (5.10) is the precondition that guarantees the periodic relative motion with no secular drifts in circular orbits.

By substituting Eq. (5.10) into Eqs. (5.7)–(5.9), the analytic solutions reduce to

$$x(t) = (\dot{x}_0/n_C)\sin n_T t + x_0 \cos n_C t \tag{5.11}$$

$$y(t) = (2\dot{x}_0/n_C)\cos n_C t - 2x_0 \sin n_C t + y_0 - 2\dot{x}_0/n_C \tag{5.12}$$

$$z(t) = (\dot{z}_0/n_C)\sin n_C t + z_0 \cos n_C t \tag{5.13}$$

Thus, the in-plane relative trajectory is an ellipse, given by

$$x^2/d^2 + (y-c)^2/(4d^2) = 1 \tag{5.14}$$

where

$$d = \sqrt{x_0^2 + (\dot{x}_0/n_C)^2}, \quad c = y_0 - 2\dot{x}_0/n_C \tag{5.15}$$

As can be seen, the center of the ellipse is located at $(0, c)$. Obvious, if and only if $c = 0$, that is,

$$y_0 - 2\dot{x}_0/n_C = 0 \tag{5.16}$$

then the center of the ellipse is located at the origin $(0, 0)$.

The common configurations used in circular reference orbits include the leader–follower (LF), projected circular orbit (PCO), general circular orbit (GCO) configurations, and so on [3]. In the LF configuration, the deputy and the chief spacecraft fly in the same circular orbit, and the deputy only leads or lags the chief by a constant phase angle. A brief introduction of the PCO and GCO configuration will be given as follows:

5.2.1.1 PCO Configuration

In PCO configuration, the projection of the relative trajectory on the yz plane is a circle with a radius of $2d$, that is, [3]

$$y^2 + z^2 = 4d^2 \tag{5.17}$$

In view of $y_0^2 + z_0^2 = 4d^2$, it can be derived that [3]

$$z_0 = \pm 2x_0 \tag{5.18}$$

$$\dot{z}_0 = \pm 2\dot{x}_0 \tag{5.19}$$

Notably, the signs in the right sides of Eqs. (5.18) and (5.19) should be positive or negative signs at the same time.

Thus, to formulate a PCO configuration with its center located at the origin of the LVLH frame, the initial relative states should satisfy Eqs. (5.10), (5.16), (5.18), and (5.19).

5.2.1.2 GCO Configuration

In GCO configuration, the relative trajectory is a circle with a radius of $2d$, given by [3],

$$x^2 + y^2 + z^2 = 4d^2 \tag{5.20}$$

Also, in view of $x_0^2 + y_0^2 + z_0^2 = 4d^2$, it holds that [3]

$$z_0 = \pm\sqrt{3}x_0 \tag{5.21}$$

$$\dot{z}_0 = \pm\sqrt{3}\dot{x}_0 \tag{5.22}$$

Similarly, the signs in the right sides of Eqs. (5.21) and (5.22) should be positive or negative signs at the same time.

Thus, to formulate a GCO configuration with its center located at the origin of the LVLH frame, the initial relative states should satisfy Eqs. (5.10), (5.16), (5.21), and (5.22).

5.2.2 *Problem Formulation*

In this section, the Lorentz-propelled spacecraft formation establishment or reconfiguration problem will be formulated as a general trajectory optimization problem (TOP). Generally, the constraints in a TOP include dynamic constraints, boundary constraints, and path constraints. A typical TOP can be formulated as follows [4]. Find the optimal profile of the state-control pair $\{\boldsymbol{x}, \boldsymbol{u}\}$ that minimizes the following Lagrange cost function:

$$J_l = \int_{t_0}^{t_f} L[\boldsymbol{x}(t), \boldsymbol{u}(t), t]\mathrm{d}t \tag{5.23}$$

where $L : \mathbb{R}^n \times \mathbb{R}^m \times \mathbb{R} \to \mathbb{R}$ is the running cost function. t_0 and t_f refer to the initial and final time, respectively. $\boldsymbol{x} \in \mathbb{R}^n$ and $\boldsymbol{u} \in \mathbb{R}^m$ are the state and control vectors subject to the following dynamic constraints:

$$\dot{\boldsymbol{x}} = \boldsymbol{f}[\boldsymbol{x}(t), \boldsymbol{u}(t), t] \tag{5.24}$$

boundary constraints

$$\boldsymbol{\varphi}_l \le \boldsymbol{\varphi}[\boldsymbol{x}(t_0), t_0, \boldsymbol{x}(t_f), t_f] \le \boldsymbol{\varphi}_r \tag{5.25}$$

and path constraints

$$\boldsymbol{\psi}_l \le \boldsymbol{\psi}[\boldsymbol{x}(t_0), t_0, \mathbf{x}(t_f), t_f] \le \boldsymbol{\psi}_r \tag{5.26}$$

where $\boldsymbol{\varphi}_l \in \mathbb{R}^p$ and $\boldsymbol{\varphi}_r \in \mathbb{R}^p$ are the lower and upper bounds of the boundary constraints. $\boldsymbol{\psi}_l \in \mathbb{R}^q$ and $\boldsymbol{\psi}_r \in \mathbb{R}^q$ are the lower and upper bounds of the path constraints.

Based on the definitions above, the Lorentz-propelled TOPs can be formulated as follows. Suppose that the specific charge of the deputy spacecraft (i.e., λ) is the single control input (i.e., $\boldsymbol{a}_C = 0$), indicating that only the Lorentz acceleration is used to perform orbital maneuvers. Find the optimal trajectory of λ and possibly the charging time t_f that minimize the following objective function:

$$J = \int_{t_0}^{t_f} \lambda^2 \mathrm{d}t \tag{5.27}$$

subject to the dynamic constraints in Eq. (5.6). Other constraints including the boundary and path constraints are summarized as follows.

5.2.2.1 Initial Condition Constraints

The initial condition constraints are just the initial relative states between the chief and deputy, given by

$$\begin{aligned} x(t_0) &= x_0, \quad y(t_0) = y_0, \quad z(t_0) = z_0 \\ \dot{x}(t_0) &= \dot{x}_0, \quad \dot{y}(t_0) = \dot{y}_0, \quad \dot{z}(t_0) = \dot{z}_0 \end{aligned} \tag{5.28}$$

5.2.2.2 Final Condition Constraints

Two kinds of final condition constraints are considered in this section, that is, the fixed and free final condition constraints [5].

For the fixed one, the terminal relative states are predetermined as fixed points that satisfy the configuration requirements of a desired formation, given by

$$\begin{aligned} x(t_f) &= x_f, \quad y(t_f) = y_f, \quad z(t_0) = z_f \\ \dot{x}(t_f) &= \dot{x}_f, \quad \dot{y}(t_f) = \dot{y}_f, \quad \dot{z}(t_0) = \dot{z}_f \end{aligned} \tag{5.29}$$

Differently, for the free one, the terminal relative states are not determined in advance, but are treated as optimization variables. Actually, the terminal relative states can be arbitrary ones as long as they satisfy the configuration requirements of the desired formation. For example, in view of the formation conditions in Sect. 5.2.1.1, to form a PCO with a radius of ρ, the free final condition constraints can be formulated as [3]

$$\dot{y}_f = -2n_C x_f \tag{5.30}$$

$$y_f - 2\dot{x}_f/n_C = 0 \tag{5.31}$$

$$2\sqrt{x_f^2 + (\dot{x}_f/n_C)^2} = \rho \tag{5.32}$$

$$z_f = \pm 2x_f \tag{5.33}$$

$$\dot{z}_f = \pm 2\dot{x}_f \tag{5.34}$$

Notably, Eq. (5.30) ensures the periodic relative motion in circular orbits without secular drifts. Equation (5.31) guarantees that the formation center is located at the origin of the LVLH frame, and Eq. (5.32) implies that the radius of the PCO is ρ. Equations (5.33) and (5.34) are the geometric requirements of a PCO formation. Likewise, it is notable that the signs in the right sides of Eqs. (5.33) and (5.34) should be positive or negative signs at the same time.

Furthermore, if a GCO formation is desired, then the conditions in Eqs. (5.33) and (5.34) should be replaced with [3]

$$z_f = \pm\sqrt{3}x_f \tag{5.35}$$

$$\dot{z}_f = \pm\sqrt{3}\dot{x}_f \tag{5.36}$$

Other three conditions remain unchanged.

5.2.2.3 Path Constraints

Given that the near-term feasible maximum of the specific charge is about 0.03 C/kg, a path constraint imposed on the control input is chosen as

$$-0.03 \le \lambda \le 0.03 \tag{5.37}$$

5.2.3 *Open-Loop Control*

Having formulated the Lorentz-propelled formation establishment and reconfiguration problem as general TOPs, the next step is to solve the TOPs via numerical method. In this section, Gauss pseudospectral method (GPM), a direct optimization method, is used to solve the TOPs. GPM transcribes a TOP into nonlinear programming (NLP) by parameterizing the state and control variables using orthogonal polynomials and approximating the dynamics at Gauss quadrature collocation points [4]. In GPM, the Legendre polynomials and Legendre–Gauss (LG) points are used. The resulting NLP is then solved by appropriate numerical methods. A brief introduction of GPM is summarized as follows [4].

Consider the optimal control problem described from Eqs. (5.23) to (5.26). GPM discretizes and transcribes the optimal control problem into an NLP. The time interval $t \in [t_0, t_f]$ can be transformed into time interval $\tau \in [-1, 1]$ via affine transformation [4]:

$$t = \frac{t_f - t_0}{2}\tau + \frac{t_f + t_0}{2} \tag{5.38}$$

GPM is developed based on Legendre–Gauss (LG) collocation points, which lie on the open interval $\tau \in (-1, 1)$. This set of points includes neither of the end points (i.e., $\tau_0 = -1$ and $\tau_f = 1$). In GPM, the state is approximated by the following Lagrange interpolating polynomials of degree $N + 1$ [4]:

$$\boldsymbol{x}(\tau) \approx \boldsymbol{X}(\tau) = \sum_{i=0}^{N} \boldsymbol{X}(\tau_i) L_i(\tau) \tag{5.39}$$

where τ_i $(i = 1, \ldots, N)$ are the LG points and $\tau_0 = -1$. $L_i(\tau)$ is defined as

$$L_i(\tau) = \prod_{j=0, j \ne i}^{N} \frac{\tau - \tau_j}{\tau_i - \tau_j}, \quad i = 0, 1, \ldots, N \tag{5.40}$$

The control is approximated using the following Lagrange interpolating polynomials of degree N [4]:

$$\boldsymbol{u}(\tau) \approx \boldsymbol{U}(\tau) = \sum_{i=1}^{N} \boldsymbol{U}(\tau_i)\tilde{L}_i(\tau) \tag{5.41}$$

where

$$\tilde{L}_i(\tau) = \prod_{j=1, j \neq i}^{N} \frac{\tau - \tau_j}{\tau_i - \tau_j}, \quad i = 1, \ldots, N \tag{5.42}$$

The derivative of state vector $\boldsymbol{x}(\tau)$, collocated at LG points $\tau_k (k = 1, \ldots, N)$, can be obtained by differentiating Eq. (5.39) as follows [4]:

$$\dot{\boldsymbol{x}}(\tau_k) \approx \dot{\boldsymbol{X}}(\tau_k) = \sum_{i=0}^{N} \boldsymbol{X}(\tau_i)\dot{L}_i(\tau) = \sum_{i=0}^{N} D_{ki}\boldsymbol{X}(\tau_i) \tag{5.43}$$

where $\boldsymbol{D} \in \mathbb{R}^{N \times (N+1)}$ is the differentiation matrix, given by

$$D_{ki} = \dot{L}_i(\tau_k) = \sum_{l=0}^{N} \frac{\prod_{j=0, j \neq i, l}^{N} (\tau_k - \tau_j)}{\prod_{j=0, j \neq i}^{N} (\tau_i - \tau_j)} \tag{5.44}$$

To summarize, the optimal control problem is therefore approximated by the following NLP. Find the coefficients $\{\boldsymbol{X}, \boldsymbol{U}\}$ and possibly the final time t_f that minimize the cost function [4]

$$J = \frac{t_f - t_0}{2} \sum_{k=1}^{N} w_k \mathcal{L}[\boldsymbol{X}(\tau_k), \boldsymbol{U}(\tau_k), \tau_k] \tag{5.45}$$

subject to

$$\sum_{i=0}^{N} D_{ki}\boldsymbol{X}(\tau_i) - \frac{t_f - t_0}{2} f[\boldsymbol{X}(\tau_k), \boldsymbol{U}(\tau_k), \tau_k] = \boldsymbol{0}, \quad k = 1, \ldots, N \tag{5.46}$$

$$\boldsymbol{\varphi}_l \leq \boldsymbol{\varphi}[\boldsymbol{X}(\tau_0), t_0, \boldsymbol{X}(\tau_f), t_f] \leq \boldsymbol{\varphi}_r \tag{5.47}$$

$$\boldsymbol{\psi}_l \leq \boldsymbol{\psi}[\boldsymbol{X}(\tau_k), \boldsymbol{U}(\tau_k), \tau_k] \leq \boldsymbol{\psi}_r, \quad k = 1, \ldots, N \tag{5.48}$$

where w_k denote the gauss weights. It should be noted that the dynamic constraint is collocated only at LG points but not at either of the boundary points. Thus, an additional constraint should be added to guarantee that the final state $\boldsymbol{X}(\tau_f)$ satisfies the dynamic constraint, given by

$$\boldsymbol{X}(\tau_f) = \boldsymbol{X}(\tau_0) + \frac{t_f - t_0}{2}\sum_{k=1}^{N} w_k f[\boldsymbol{X}(\tau_k), \boldsymbol{U}(\tau_k), \tau_k] \quad (5.49)$$

Finally, by solving the above NLP via numerical methods, the optimal trajectories of the state and control variables can be derived.

5.2.4 *Closed-Loop Control*

As the application of Lorentz-augmented spacecraft hovering in Sect. 4.2.2, the closed-loop control schemes for Lorentz-augmented SFF in elliptic orbits will also be applicable to the case with circular orbits. Thus, given that the closed-loop control schemes for elliptic orbits will be elaborated in the following Sect. 5.3.3, discussions for the case with circular orbits are not presented separately for brevity. The reader is referred to Sect. 5.3.3 for details.

5.2.5 *Numerical Simulation*

The chief spacecraft is assumed to be flying in a circular LEO with an orbital period of 1.6 h, and its initial orbit elements are listed in Table 5.1. The initial phase angle of the dipole is set as $\Omega_0 = -60°$.

5.2.5.1 Formation Establishment

It is assumed that the deputy Lorentz spacecraft starts from the origin of the LVLH frame with zero relative velocity initially. Thus, the initial condition constraints (i.e., the initial relative states) are given by

Table 5.1 Initial orbit elements of the chief spacecraft

Orbit element	Value
Semimajor axis (km)	6945.034
Eccentricity	0
Inclination (deg)	30
Right ascension of ascending node (deg)	50
Argument of latitude (deg)	0

Reprinted from Ref. [5], Copyright 2014, with permission from Elsevier

$$\begin{array}{lll} x_0 = 0 \text{ m}, & y_0 = 0 \text{ m}, & z_0 = 0 \text{ m} \\ \dot{x}_0 = 0 \text{ m/s}, & \dot{y}_0 = 0 \text{ m/s}, & \dot{z}_0 = 0 \text{ m/s} \end{array} \tag{5.50}$$

The initial time is set as $t_0 = 0$ s, and it is required to establish a PCO formation with a radius of $\rho = 500$ m at the final time (i.e., the charge time) satisfying $t_f \leq 6000$ s. Two kinds of final condition constraints are considered. For the fixed one, the terminal relative states (i.e., the final condition constraints) are designated as

$$\begin{array}{lll} x_f = 0 \text{ m}, & y_f = 500 \text{ m}, & z_f = 0 \text{ m} \\ \dot{x}_f = 0.2727 \text{ m/s}, & \dot{y}_f = 0 \text{ m/s}, & \dot{z}_f = 0.5454 \text{ m/s} \end{array} \tag{5.51}$$

For the free one, the final condition constraints are shown in Eqs. (5.30)–(5.34). Notably, the signs in Eqs. (5.33) and (5.34) are assumed to be positive.

60 LG points are used in the GPM to solve these two kinds of TOPs. For both fixed and free condition constraints, time histories of the specific charge necessary for formation establishment are compared in Fig. 5.1. The acronym "FC" refers to final conditions. For the fixed one, the energy-optimal charge time is $t_f = 4319.3$ s, and the corresponding energy consumption is $J = 4.50 \times 10^{-2}$. For the free one, the charge time is $t_f = 6000.0$ s, and the corresponding energy consumption is $J = 3.26 \times 10^{-2}$, with nearly 27.6 % of the control energy saved. Furthermore, the optimal terminal relative states are derived as

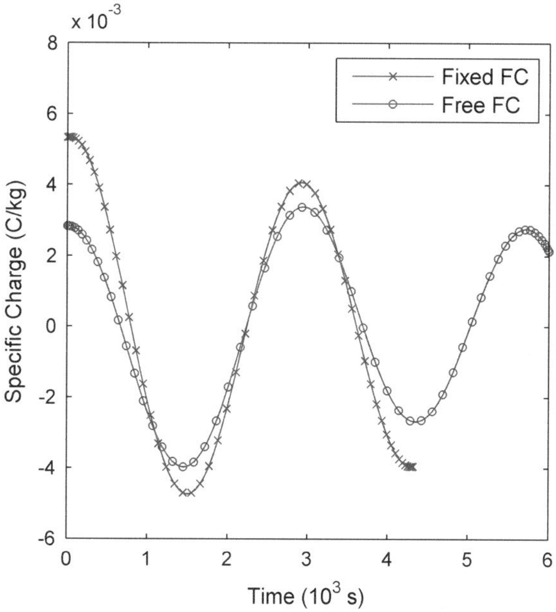

Fig. 5.1 Comparisons of the time histories of required specific charges for formation establishment. Reprinted from Ref. [5], Copyright 2014, with permission from Elsevier

$$\begin{array}{lll} x_f = 241.43\ \text{m}, & y_f = -129.79\ \text{m}, & z_f = 482.86\ \text{m} \\ \dot{x}_f = -0.0708\ \text{m/s}, & \dot{y}_f = -0.5267\ \text{m/s}, & \dot{z}_f = -0.1416\ \text{m/s} \end{array} \tag{5.52}$$

By substituting Eq. (5.52) into Eqs. (5.30)–(5.34), it can be verified that the derived terminal relative states satisfy the configuration requirements of a PCO with a radius of 500 m. Also, as shown in Fig. 5.1, the necessary charging levels for both final condition constraints are in the order of 10^{-3} C/kg, smaller than the bound of 0.03 C/kg. Moreover, the discrete points in Fig. 5.1 denote the results generated via GPM, and the solid lines denote the results of Lagrangian interpolation between LG points.

For these two kinds of final condition constraints, time histories of the optimal relative position and velocity trajectories are compared in Figs. 5.2 and 5.3, respectively. Likewise, the discrete points in both figures refer to the results derived via GPM. To verify the validity of GPM's results, the interpolated trajectory of the optimal specific charge is substituted into the dynamical model [i.e., Eq. (5.6)] to derive the real trajectories of relative position and velocity by integrating the differential equations using the fourth-order Runge–Kutta method. The integrated results are shown by the solid lines in Figs. 5.2 and 5.3.

As can be seen, the GPM's results are nearly coincided with the numerical ones, verifying the validity of GPM. For the case with fixed final condition constraints, the terminal relative position errors are 0.11 m, 0.09 m, and 0.02 m on each axis, and the terminal relative velocity errors are 7.89×10^{-5}, 2.32×10^{-4}, and 3.30×10^{-5} m/s on each axis. Similarly, for the case with free ones, the terminal relative position errors are, respectively, 0.02 m, 0.35 m, and 0.02 m. Also, the

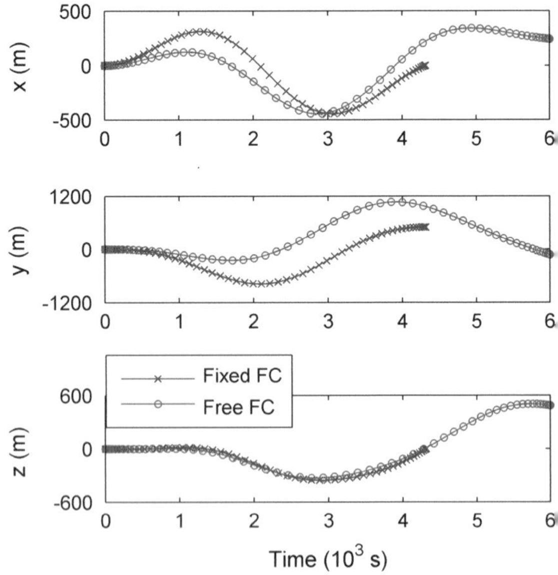

Fig. 5.2 Comparisons of the time histories of relative position for formation establishment. Reprinted from Ref. [5], Copyright 2014, with permission from Elsevier

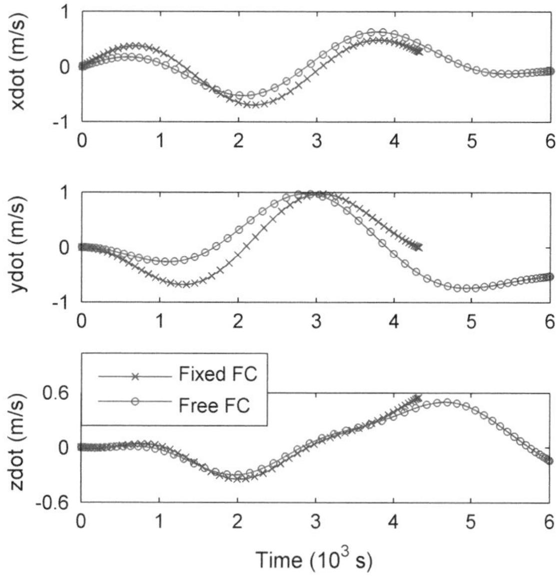

Fig. 5.3 Comparisons of the time histories of relative velocity for formation establishment. Reprinted from Ref. [5], Copyright 2014, with permission from Elsevier

terminal relative velocity errors are 1.16×10^{-4}, 4.12×10^{-5}, and 1.25×10^{-4} m/s, respectively. As can be seen, these errors are all within acceptable tolerances. These errors mainly arise from the interpolation and integration methods. Increasing the number of LG points can further reduce these errors.

The relative transfer trajectories for formation establishment of both cases are compared in Fig. 5.4. As can be seen, in both cases, the deputy departures from the origin of the LVLH frame, and establishes the same PCO at different terminal

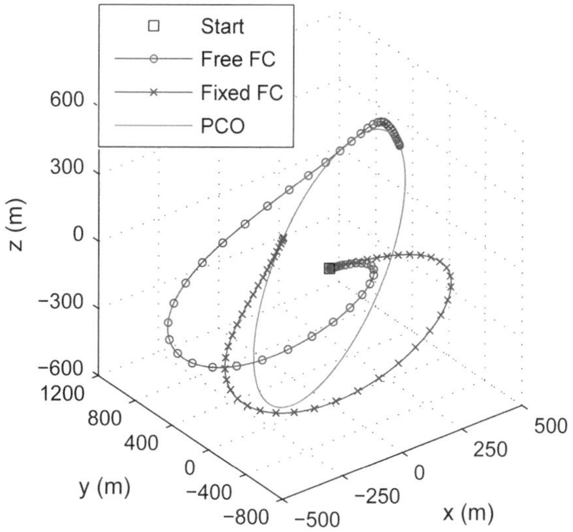

Fig. 5.4 Comparisons of the relative trajectories for formation establishment. Reprinted from Ref. [5], Copyright 2014, with permission from Elsevier

positions finally. The trajectory of PCO is derived via propagating the terminal relative states using the HCW equations. Obviously, the feasibility and validity of GPM in solving Lorentz-propelled formation establishment problem have been substantiated via this numerical example.

5.2.5.2 Formation Reconfiguration

The initial orbit elements of the chief spacecraft are set the same as those given in Table 5.1, and the initial phase angle of the dipole is also set as $\Omega_0 = -60°$.

It is assumed that the chief and deputy are in a PCO formation with a radius of $\rho = 500$ m initially, and the initial relative states are

$$\begin{aligned} x_0 &= 0\ \text{m}, & y_0 &= 500\ \text{m}, & z_0 &= 0\ \text{m} \\ \dot{x}_0 &= 0.2727\ \text{m/s}, & \dot{y}_0 &= 0\ \text{m/s}, & \dot{z}_0 &= 0.5454\ \text{m/s} \end{aligned} \tag{5.53}$$

It is required to reconfigure a GCO with a radius of $\rho = 1000$ m at the terminal time t_f satisfying $t_f \leq 6000$ s. Also, two kinds of final condition constraints are considered. For the case with fixed final condition constraints, the terminal relative states are designated in advance as

$$\begin{aligned} x_f &= 0\ \text{m}, & y_f &= -1000\ \text{m}, & z_f &= 0\ \text{m} \\ \dot{x}_f &= -0.5454\ \text{m/s}, & \dot{y}_f &= 0\ \text{m/s}, & \dot{z}_f &= 0.9447\ \text{m/s} \end{aligned} \tag{5.54}$$

For the case with free final condition constraints, the final condition constraints are given in Eqs. (5.30)–(5.32), and Eqs. (5.35) and (5.36). Notably, in this example, the signs in Eqs. (5.35) and (5.36) are assumed to be negative.

Also, 60 LG points are used in GPM to solve the TOP. For these two cases, time histories of the specific charge necessary for formation reconfiguration from PCO to GCO are compared in Fig. 5.5. Likewise, the discrete points denote the results obtained by GPM, and the solid lines refer to the Lagrangian interpolation results between the LG points. As can be seen, for both cases, the necessary charging level is in the order of 10^{-3} C/kg, smaller than the near-term feasible maximum of 0.03 C/kg. For the case with fixed final conditions, the energy-optimal charging time is derived as $t_f = 4883.4$ s, and the corresponding energy consumption is $J = 8.14 \times 10^{-2}$. Differently, for the case with free final conditions, the charging time increases to $t_f = 6000.0$ s, but the energy consumption decreases to $J = 7.38 \times 10^{-2}$, with nearly 9.34 % of the control energy saved. Furthermore, for this case, the optimal terminal relative states are derived as

$$\begin{aligned} x_f &= -469.20\ \text{m}, & y_f &= -345.54\ \text{m}, & z_f &= 812.68\ \text{m} \\ \dot{x}_f &= -0.1885\ \text{m/s}, & \dot{y}_f &= 1.0236\ \text{m/s}, & \dot{z}_f &= 0.3264\ \text{m/s} \end{aligned} \tag{5.55}$$

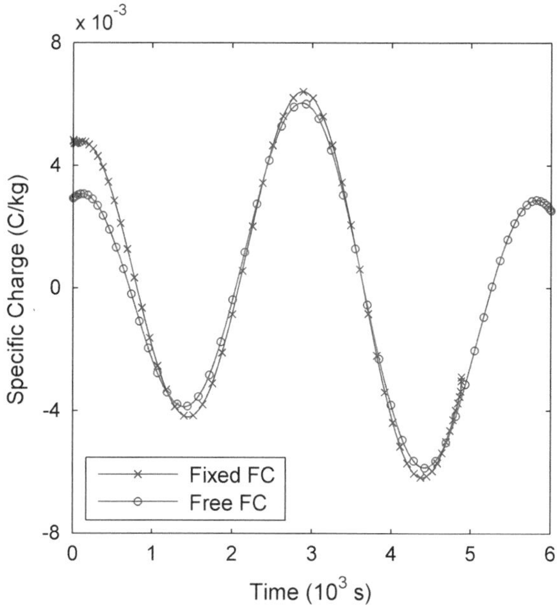

Fig. 5.5 Comparisons of the time histories of required specific charges for PCO to GCO formation reconfiguration. Reprinted from Ref. [5], Copyright 2014, with permission from Elsevier

Similarly, it can easily be verified that Eq. (5.55) satisfies the configuration requirements for a GCO with a radius of 1000 m, proving the validity of the results derived via GPM.

Relative transfer trajectories from PCO to GCO are compared in Fig. 5.6. Also, the discrete points denote the results generated via GPM. To verify the validity of the GPM's results, the interpolated optimal trajectories of the specific charge are also substituted into the dynamical model [i.e., Eq. (5.6)] to generate the real trajectories of relative position and velocity by numerical integration of the dynamic equations via fourth-order Runge–Kutta method. The numerical results are denoted by the solid lines in Fig. 5.6. As can be seen, the numerical results are nearly coincided with the GPM's results, verifying the validity of GPM. Specifically, for the case with fixed final conditions, the terminal relative position errors between the numerical and GPM's results are, respectively, 1.65×10^{-3}, 5.20×10^{-3}, and 5.20×10^{-3} m on each axis. Similarly, for the case with free final conditions, the relative position errors are 1.37×10^{-2}, 1.94×10^{-2}, and 1.01×10^{-2} m, respectively. All errors are within acceptable tolerances, which can be further reduced by choosing more LG points or using more precise integration method. Meanwhile, as shown in Fig. 5.6, for both cases, the deputy starts from the smaller PCO and reaches to the same larger GCO at different terminal positions, verifying again the feasibility and validity of GPM in solving Lorentz-propelled formation reconfiguration problem.

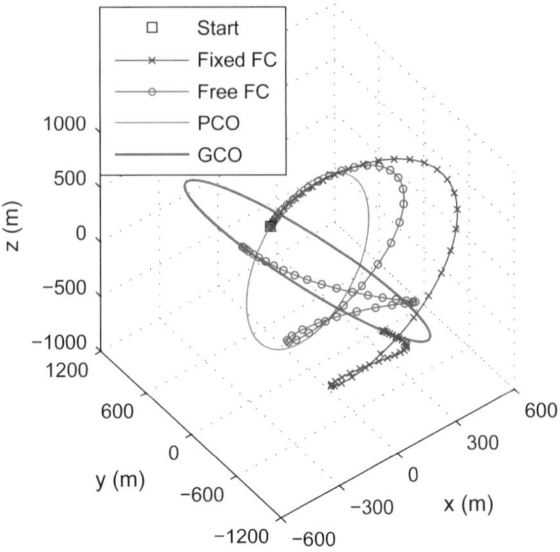

Fig. 5.6 Comparisons of the relative trajectories for PCO to GCO formation reconfiguration. Reprinted from Ref. [5], Copyright 2014, with permission from Elsevier

5.2.5.3 Effect of the Dipole Tilt Angle

In most of the previous works on Lorentz-augmented SFF, a nontilted dipole model is used. Differently, a tilted dipole model, which is more representative of the Earth's magnetic field, is introduced in this book. Thus, the effect of the dipole tilt angle on the optimal trajectories is numerically investigated here by comparisons of the results derived via the nontilted and tilted dipole models.

The formation reconfiguration from PCO to GCO with free final condition constraints, as presented in Sect. 5.2.5.2, is considered here. In the above section, the optimal results for the tilted dipole model have been derived. The required specific charge is shown in Fig. 5.5, and the relative transfer trajectory is shown in Fig. 5.6. To derive the optimal specific charge and the corresponding relative transfer trajectory for a nontilted dipole model, the dipole tilt angle is set as $\alpha = 0$ while other parameters remain unchanged.

Time histories of the specific charge necessary for reconfiguration are then compared in Fig. 5.7. For the tilted dipole model, the required specific charge is denoted as λ, and for the nontilted one, it is denoted as λ^{non}. As can be seen, despite that λ and λ^{non} are both in the order of 10^{-3} C/kg, distinct errors exist between these two models. To further investigate the effect of the dipole tilt angle, the specific charge derived in the nontilted dipole model (i.e., λ^{non}) is then substituted into the tilted dipole model to generate the trajectories of relative states. The terminal relative states are then derived as

Fig. 5.7 Comparisons of the time histories of required specific charges for formation reconfiguration in both tilted and nontilted dipole models. Reprinted from Ref. [5], Copyright 2014, with permission from Elsevier

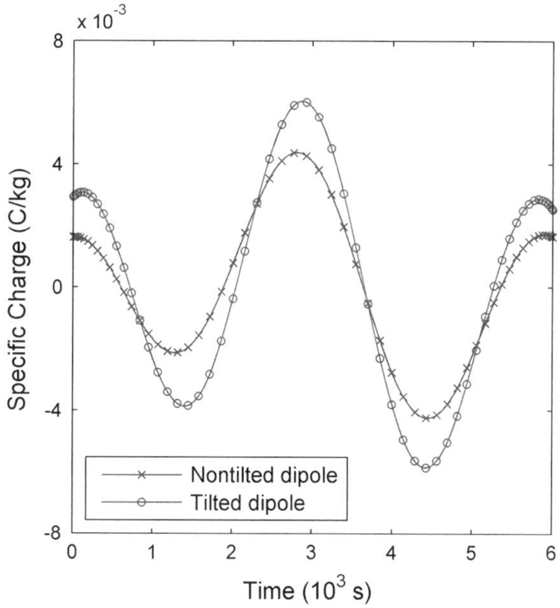

$$\begin{aligned} x_f &= -519.15\ \text{m}, & y_f &= -430.18\ \text{m}, & z_f &= 550.20\ \text{m}, \\ \dot{x}_f &= -0.2276\ \text{m/s}, & \dot{y}_f &= 1.1342\ \text{m/s}, & \dot{z}_f &= 0.3573\ \text{m/s} \end{aligned} \quad (5.56)$$

Obviously, by substituting Eq. (5.56) into Eqs. (5.30)–(5.32), Eqs. (5.35), and (5.36), it can be verified that the terminal relative states do not satisfy the configuration requirements for a GCO with a radius of 1000 m, indicating the failure of the reconfiguration mission.

Furthermore, to show the mission failure more distinctively, Fig. 5.8 depicts the relative trajectories propagating from the terminal relative states using the HCW equation. The total simulation time is set as 3×10^4 s, about five orbital periods. As shown in Fig. 5.8, because λ^{non} is derived in the nontilted dipole model, then the GCO formation will be reconfigured if λ^{non} is inputted in the nontilted dipole model. However, if λ^{non} is inputted in the tilted dipole method, then the GCO formation will not be reconfigured any more, and obvious offsets exist after each period. Therefore, the dipole tilt angle of the Earth's magnetic field should be considered when analyzing the Lorentz-propelled formation establishment or reconfiguration problem in Earth orbits.

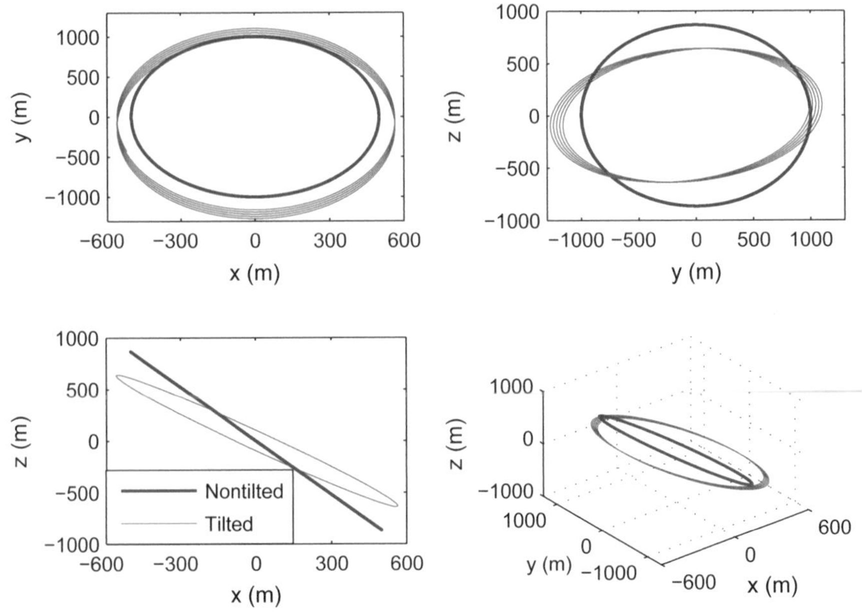

Fig. 5.8 Comparisons of propagated relative trajectories in both nontilted and tilted dipole models. Reprinted from Ref. [5], Copyright 2014, with permission from Elsevier

5.3 Case of Two-Body Elliptic Reference Orbit

In this section, the chief is assumed to be flying in an elliptic orbit. Thus, the dynamic equations of Lorentz-augmented SFF are given in Eq. (5.1). As an elliptic Keplerian orbit, the orbital motion of the chief spacecraft is governed by [6]

$$\begin{aligned} \ddot{R}_C &= R_C \dot{u}_C^2 - \mu / R_C^2 \\ R_C \ddot{u}_C &= -2 \dot{R}_C \dot{u}_C \end{aligned} \tag{5.57}$$

5.3.1 Formation Configurations in Elliptic Orbits

Consider a chief spacecraft in an elliptic orbit. The orbital motion of the deputy relative to the chief is periodic if and only if the semimajor axis or the orbital period of the deputy is exactly equal to that of the chief's orbit. Note that the semimajor axis corresponds the orbital energy, that is, $E = -\mu/2a = V^2/2 - \mu/R$, where E is the orbital energy, a is the semimajor axis, V is the velocity, and R is the orbital radius of the spacecraft. Thus, the aforementioned periodicity condition can be formulated as the following energy-matching condition [7]:

$$E_C = -\frac{\mu}{2a_C} = E_L = \frac{1}{2}V_L^2 - \frac{\mu}{R_L} \tag{5.58}$$

where a_C is the semimajor axis of the chief's orbit. E_C and E_L refer to the orbital energy of the chief and the deputy, respectively. V_L and R_L denote the velocity and orbital radius of the deputy Lorentz spacecraft, respectively.

Notably, the condition in Eq. (5.58) can be normalized and then represented by the relative states as

$$(\dot{\tilde{R}}_C + \dot{\tilde{x}} - \tilde{y}\tilde{\dot{u}}_C)^2 + [\dot{\tilde{y}} + (\tilde{R}_C + \tilde{x})\tilde{\dot{u}}_C]^2 + \dot{\tilde{z}}^2 - 2/\sqrt{(\tilde{R}_C + \tilde{x})^2 + \tilde{y}^2 + \tilde{z}^2} = -1 \tag{5.59}$$

where the superscript $\sim$ refers to a normalized variable. The position is normalized by a_C, the velocity is normalized by $\sqrt{\mu/a_C}$, and the angular velocity is normalized by $\sqrt{\mu/a_C^3}$.

In other words, if the relative states between the chief and the deputy satisfy the energy-matching condition in Eq. (5.59), then the relative orbits between these two Keplerian orbits are bounded and periodic ones.

5.3.2 *Open-Loop Control*

Similar to the case with a circular reference orbit, the optimal control problem of Lorentz-augmented spacecraft formation establishment or reconfiguration in elliptic orbits is first formulated as a general TOP. Different from the case with circular reference orbit where only Lorentz acceleration is used for propulsion, a hybrid control input consisting of the specific charge and thruster-generated control acceleration is considered for the case with elliptic reference orbit. Define the hybrid control input as $\boldsymbol{U} = [\lambda \quad a_R \quad a_S \quad a_W]^{\mathrm{T}}$, and define the relative state vector as $\boldsymbol{X} = [\boldsymbol{\rho}^{\mathrm{T}} \quad \dot{\boldsymbol{\rho}}^{\mathrm{T}}]^{\mathrm{T}}$. Then, the TOP can be formulated as follows. Find the optimal profiles of $\{\boldsymbol{X}, \boldsymbol{U}\}$ that minimizes the following objective function:

$$J = \int_{t_0}^{t_f} \boldsymbol{U}^{\mathrm{T}}\boldsymbol{Q}\boldsymbol{U}\mathrm{d}t \tag{5.60}$$

where $\boldsymbol{Q}$ is a weight matrix, given by

$$\boldsymbol{Q} = \mathrm{diag}(Q_q, Q_u, Q_u, Q_u), \quad (Q_q > 0, Q_u > 0) \tag{5.61}$$

The dynamic constraints have been given in Eqs. (5.1) and (5.57). The boundary and path constraints are summarized as follows:

The boundary constraints consist of the initial and final condition constraints. Similarly, the initial condition constraints are just the initial relative states between the chief and the deputy, given by

$$\boldsymbol{X}(t_0) = \begin{bmatrix} x_0 & y_0 & z_0 & \dot{x}_0 & \dot{y}_0 & \dot{z}_0 \end{bmatrix}^{\mathrm{T}} \tag{5.62}$$

Notably, only fixed final condition constraints are considered in this section. Thus, to establish or reconfigure a desired formation configuration in elliptic orbits, the final condition constraints are the predetermined condition constraints that satisfy the energy-matching conditions in Eq. (5.59), given by

$$\boldsymbol{X}(t_f) = \begin{bmatrix} x_f & y_f & z_f & \dot{x}_f & \dot{y}_f & \dot{z}_f \end{bmatrix}^{\mathrm{T}} \tag{5.63}$$

Also, given that the near-term feasible maximal specific charge is about 0.03 C/kg, a path constraint imposed on the specific charge is proposed as

$$-0.03 \le \lambda \le 0.03 \tag{5.64}$$

Having formulated the above TOP, the next step is to solve this TOP using GPM as presented in Sect. 5.2.3. In this way, the optimal trajectories of Lorentz-augmented formation establishment or reconfiguration in elliptic orbits can be derived.

Remark 5.1 The values of Q_q and Q_u determine the weights on these two kinds of propulsion. Obviously, increasing Q_u and decreasing Q_q could reduce fuel consumptions. Given that propellant is a limited resource on board, fuel consumption should be given priority. If $Q_q \ll Q_u$, then the thruster-generated control acceleration necessary for formation establishment or reconfiguration will be considerably small as compared to the Lorentz acceleration, that is, $\boldsymbol{a}_C \ll \boldsymbol{a}_L$. If this is the case, the resulting control acceleration $\boldsymbol{a}_C$ can be regarded as external perturbations, and a nearly propellantless formation establishment or reconfiguration is achieved. However, if the value of Q_q is too small, then the required charging level may exceed the maximum of 0.03 C/kg. Thus, the weight matrix $\boldsymbol{Q}$ is chosen to find a balance between the fuel consumption and the allowed charging level of the Lorentz spacecraft.

5.3.3 Closed-Loop Control

Considering that the external perturbations that may perturb the desired open-loop trajectory (i.e., the optimal one generated via GPM), the perturbed dynamical model of Lorentz-augmented SFF can be revised from Eq. (5.1) as

$$\ddot{\boldsymbol{\rho}} = \boldsymbol{M}(\boldsymbol{\rho}, \dot{\boldsymbol{\rho}}) + \boldsymbol{u} + \boldsymbol{d} \tag{5.65}$$

with

$$\boldsymbol{M}(\boldsymbol{\rho}, \dot{\boldsymbol{\rho}}) = \begin{bmatrix} 2\dot{u}_C\dot{y} + \dot{u}_C^2 x + \ddot{u}_C y + n_C^2 R_C - n_L^2(R_C + x) \\ -2\dot{u}_C\dot{x} + \dot{u}_C^2 y - \ddot{u}_C x - n_L^2 y \\ -n_L^2 z \end{bmatrix} \tag{5.66}$$

where $\boldsymbol{u} = \boldsymbol{a}_L + \boldsymbol{a}_C$ is the total acceleration acting on the Lorentz spacecraft, $\boldsymbol{d} = [d_x \quad d_y \quad d_z]^{\mathrm{T}}$ refers to the bounded external perturbations, satisfying $|d_i| \le d_m$, $(i = x, y, z)$.

Define $\boldsymbol{\rho}_d$ as the desired relative position, then the relative position error can be defined as $\boldsymbol{e} = \boldsymbol{\rho} - \boldsymbol{\rho}_d = [e_x \quad e_y \quad e_z]^{\mathrm{T}}$. Likewise, the relative velocity error can be defined as $\dot{\boldsymbol{e}} = \dot{\boldsymbol{\rho}} - \dot{\boldsymbol{\rho}}_d = [\dot{e}_x \quad \dot{e}_y \quad \dot{e}_z]^{\mathrm{T}}$. Thus, the error dynamical can be represented as

$$\ddot{\boldsymbol{e}} = -\ddot{\boldsymbol{\rho}}_d + \boldsymbol{M}(\boldsymbol{\rho}, \dot{\boldsymbol{\rho}}) + \boldsymbol{u} + \boldsymbol{d} \tag{5.67}$$

with

$$\ddot{\boldsymbol{\rho}}_d = \boldsymbol{M}(\boldsymbol{\rho}_d, \dot{\boldsymbol{\rho}}_d) + \lambda_d \boldsymbol{l}_d + \boldsymbol{a}_{Cd} \tag{5.68}$$

where λ_d and $\boldsymbol{a}_{Cd}$ are the optimal control inputs derived via GPM.

The SMC design generally begins with the selection of a sliding surface. Thus, a fast nonsingular terminal sliding surface is designed as [8]

$$\boldsymbol{s} = \dot{\boldsymbol{e}} + \boldsymbol{\zeta}_1 \boldsymbol{e} + \boldsymbol{\zeta}_2 \boldsymbol{\xi}(\boldsymbol{e}) \tag{5.69}$$

with

$$\xi_i(e_i) = \begin{cases} e_i^{q/p} & \bar{s}_i = 0 \quad \text{or} \quad \bar{s}_i \ne 0, |e_i| \ge \delta \\ \nu_{1i} e_i + \nu_{2i} \operatorname{sgn}(e_i) e_i^2, & \bar{s}_i \ne 0, |e_i| < \delta \end{cases} \tag{5.70}$$

where $i = x, y, z$ and $j = 1, 2$. $\boldsymbol{\zeta}_j = \operatorname{diag}(\zeta_{jx}, \zeta_{jy}, \zeta_{jz})$ is a diagonal positive definite matrix, and $\boldsymbol{\xi}(\boldsymbol{e}) = [\xi_x(e_x) \quad \xi_y(e_y) \quad \xi_z(e_z)]^{\mathrm{T}}$. p and q are odd integers that satisfy $1/2 < q/p < 1$. To ensure the continuity of $\xi_i(e_i)$ and its time derivative at the switch point $|e_i| = \delta$, ν_{1i} and ν_{2i} are designed as $\nu_{1i} = (2 - q/p)\delta^{q/p-1}$ and $\nu_{2i} = (q/p - 1)\delta^{q/p-2}$, respectively. δ is a small positive constant. Furthermore, $\bar{s}_i$ is defined as $\bar{s}_i = \dot{e}_i + \zeta_{1i} e_i + \zeta_{2i} e_i^{q/p}$.

Taking the time derivative of the sliding surface yields that

$$\dot{\boldsymbol{s}} = \ddot{\boldsymbol{e}} + \zeta_1 \dot{\boldsymbol{e}} + \zeta_2 \dot{\boldsymbol{\xi}}(\boldsymbol{e}, \dot{\boldsymbol{e}}) \tag{5.71}$$

with

$$\dot{\xi}_i(e_i, \dot{e}_i) = \begin{cases} (q/p) e_i^{q/p-1} \dot{e}_i & \bar{s}_i = 0 \quad \text{or} \quad \bar{s}_i \neq 0, |e_i| \geq \delta \\ \nu_{1i} \dot{e}_i + 2\nu_{2i} \mathrm{sgn}(e_i) e_i \dot{e}_i, & \bar{s}_i \neq 0, |e_i| < \delta \end{cases} \tag{5.72}$$

where $i = x, y, z$.

Then, by evaluation of $\dot{\boldsymbol{s}} = \boldsymbol{0}$, the equivalent control $\boldsymbol{u}_{\mathrm{eq}}$ can be derived as

$$\boldsymbol{u}_{\mathrm{eq}} = \ddot{\boldsymbol{\rho}}_d - \boldsymbol{M}(\boldsymbol{\rho}, \dot{\boldsymbol{\rho}}) - \zeta_1 \dot{\boldsymbol{e}} - \zeta_2 \dot{\boldsymbol{\xi}}(\boldsymbol{e}, \dot{\boldsymbol{e}}) \tag{5.73}$$

A continuous reaching law is chosen as [9]

$$\dot{\boldsymbol{s}} = -\boldsymbol{K}_1 \boldsymbol{s} - \boldsymbol{K}_2 \mathrm{sig}^{\gamma}(\boldsymbol{s}) \tag{5.74}$$

where $\boldsymbol{K}_j = \mathrm{diag}(K_{jx}, K_{jy}, K_{jz})$ $(j = 1, 2)$ is a diagonal positive definite matrix. $\mathrm{sig}^{\gamma}(\boldsymbol{s}) = \left[\, |s_x|^{\gamma} \mathrm{sgn}(s_x) \quad |s_y|^{\gamma} \mathrm{sgn}(s_y) \quad |s_z|^{\gamma} \mathrm{sgn}(s_z) \,\right]^{\mathrm{T}}$, where $0 < \gamma < 1$ is a design constant.

Thus, the fast nonsingular terminal sliding mode controller (FNTSMC) is supplemented to

$$\boldsymbol{u} = \ddot{\boldsymbol{\rho}}_d - \boldsymbol{M}(\boldsymbol{\rho}, \dot{\boldsymbol{\rho}}) - \zeta_1 \dot{\boldsymbol{e}} - \zeta_2 \dot{\boldsymbol{\xi}}(\boldsymbol{e}, \dot{\boldsymbol{e}}) - \boldsymbol{K}_1 \boldsymbol{s} - \boldsymbol{K}_2 \mathrm{sig}^{\gamma}(\boldsymbol{s}) \tag{5.75}$$

Notably, the total control acceleration $\boldsymbol{u}$ is provided by the Lorentz acceleration and the thruster-generated control acceleration together. Therefore, it remains to distribute these two kinds of propulsion in a fuel optimal way. Consider the following objective function:

$$J_d = \int_{t_0}^{t_f} L[\lambda(t), t] \mathrm{d}t = \int_{t_0}^{t_f} \|\boldsymbol{a}_C\| \mathrm{d}t = \int_{t_0}^{t_f} \|\boldsymbol{u} - \lambda \boldsymbol{l}\| \mathrm{d}t \tag{5.76}$$

Solving the Euler–Lagrange equation

$$\frac{\mathrm{d}}{\mathrm{d}t}\left(\frac{\partial L}{\partial \dot{\lambda}}\right) - \frac{\partial L}{\partial \lambda} = 0 \tag{5.77}$$

yields the optimal trajectory of the specific charge, given by

$$\lambda^* = \begin{cases} (\boldsymbol{u} \cdot \boldsymbol{l})/||\boldsymbol{l}||^2, & ||\boldsymbol{l}|| \neq 0 \\ 0, & ||\boldsymbol{l}|| = 0 \end{cases} \tag{5.78}$$

Then, the optimal trajectory of the thruster-generated control acceleration is obtained as

$$\boldsymbol{a}_C^* = \begin{cases} \boldsymbol{u} - [(\boldsymbol{u} \cdot \boldsymbol{l})/||\boldsymbol{l}||^2]\boldsymbol{l}, & ||\boldsymbol{l}|| \neq 0 \\ \boldsymbol{u}, & ||\boldsymbol{l}|| = 0 \end{cases} \tag{5.79}$$

Note that the near-term feasible maximal specific charge is about $\lambda_m = 0.03$ C/kg. Thus, if the required charging level exceeds the maximum, the control law is revised as

$$\begin{aligned} \lambda^* &= \lambda_m \mathrm{sgn}(\lambda^*) \\ \boldsymbol{a}_C^* &= \boldsymbol{u} - \lambda^* \boldsymbol{l} \end{aligned} \tag{5.80}$$

Before the stability analysis of the closed-loop system, some useful lemmas are introduced at first.

Lemma 5.1 [9]. *If* $a_1, a_2, \ldots, a_n$ *and* $0 < \kappa < 2$ are *positive, then it holds that*

$$\left(\sum_{i=1}^{n} a_i^2\right)^{\kappa} \leq \left(\sum_{i=1}^{n} a_i^{\kappa}\right)^2 \tag{5.81}$$

Lemma 5.2 [9]. *If a positive definite function* V *satisfies the inequality*

$$\dot{V} + k_1 V + k_2 V^{k_3} \leq 0 \tag{5.82}$$

where k_1, k_2, *and* k_3 *are positive constants with* $0 < k_3 < 1$, *then* V *will converge to the equilibrium in finite time* t_k *as*

$$t_k \leq \frac{1}{k_1(1-k_3)} \ln \frac{k_1 V^{1-k_3}(t_0) + k_2}{k_2} \tag{5.83}$$

Lemma 5.3 [10]. *If the system trajectory reaches the fast nonsingular terminal sliding surface* (i.e., Eq. (5.69) *with* $\delta = 0$) *at time* t_r, *then the system errors will converge to zero in finite time* t_s *as*

$$t_s = \max_{i=x,y,z} \left\{ \frac{p}{\zeta_{1i}(p-q)} \ln \frac{\zeta_{1i} e_x^{(p-q)/q}(t_r) + \zeta_{2i}}{\zeta_{2i}} \right\} \tag{5.84}$$

Now, the stability analysis of the closed-loop system is summarized in the following theorem.

Theorem 5.1 *For the dynamical system of Lorentz-augmented SFF in* Eq. (5.67), *if the sliding surface is chosen as* Eq. (5.69), *and the control laws are designed as* Eqs. (5.78) to (5.80), *then the system trajectory will converge to the neighborhood of* $\boldsymbol{s} = \boldsymbol{0}$ as

$$|s_i| \leq \varphi \tag{5.85}$$

with

$$\varphi = \min\left\{ d_m / K_{1\min}, (d_m / K_{2\min})^{1/\gamma} \right\} \tag{5.86}$$

in finite time, where $K_{j\min}$ $(j = 1, 2)$ *is the minimum eigenvalue of the* matrix $\boldsymbol{K}_j$. *Thereafter, the system errors will converge in finite time to the regions as*

$$|e_i| \leq \varphi_{ei} = \max\left\{ \delta, \min\left\{ \varphi/\zeta_{1i}, (\varphi/\zeta_{2i})^{p/q} \right\} \right\}, \quad (i = x, y, z) \tag{5.87}$$

and

$$|\dot{e}_i| \leq \varphi_{\dot{e}i} = \varphi + \zeta_{1i}\varphi_{ei} + \zeta_{2i}\varphi_{ei}^{q/p}, \quad (i = x, y, z) \tag{5.88}$$

Proof Consider the Lyapunov candidate $V = (1/2)\boldsymbol{s}^{\mathrm{T}}\boldsymbol{s} > 0$, $\forall \boldsymbol{s} \neq \boldsymbol{0}$. Taking the time derivative of V along the system trajectory yields that

$$\dot{V} = \boldsymbol{s}^{\mathrm{T}}\dot{\boldsymbol{s}} = \boldsymbol{s}^{\mathrm{T}}[\ddot{e} + \zeta_1\dot{\boldsymbol{e}} + \zeta_2\dot{\xi}(\boldsymbol{e}, \dot{\boldsymbol{e}})] \tag{5.89}$$

Substitution of Eqs. (5.67) and (5.75) into Eq. (5.89) yields that

$$\begin{aligned} \dot{V} &= \boldsymbol{s}^{\mathrm{T}}[-\ddot{\boldsymbol{\rho}}_d + \boldsymbol{M}(\boldsymbol{\rho}, \dot{\boldsymbol{\rho}}) + \boldsymbol{u} + \boldsymbol{d} + \zeta_1\dot{\boldsymbol{e}} + \zeta_2\dot{\xi}(\boldsymbol{e}, \dot{\boldsymbol{e}})] \\ &= \boldsymbol{s}^{\mathrm{T}}[\boldsymbol{d} - \boldsymbol{K}_1\boldsymbol{s} - \boldsymbol{K}_2\mathrm{sig}^{\gamma}(\boldsymbol{s})] \\ &= -\boldsymbol{s}^{\mathrm{T}}(\boldsymbol{K}_1 - \tilde{\boldsymbol{D}}\tilde{\boldsymbol{S}}^{-1})\boldsymbol{s} - \boldsymbol{s}^{\mathrm{T}}\boldsymbol{K}_2\mathrm{sig}^{\gamma}(\boldsymbol{s}) \\ &\leq -\boldsymbol{s}^{\mathrm{T}}(\boldsymbol{K}_1 - |\tilde{\boldsymbol{D}}||\tilde{\boldsymbol{S}}|^{-1})\boldsymbol{s} - \boldsymbol{s}^{\mathrm{T}}\boldsymbol{K}_2\mathrm{sig}^{\gamma}(\boldsymbol{s}) \end{aligned} \tag{5.90}$$

where $\tilde{\boldsymbol{D}} = \mathrm{diag}(d_x, d_y, d_z)$ and $\tilde{\boldsymbol{S}} = \mathrm{diag}(s_x, s_y, s_z)$ are both diagonal matrices. Also, $|\tilde{\boldsymbol{D}}| = \mathrm{diag}(|d_x|, |d_y|, |d_z|)$ and $|\tilde{\boldsymbol{S}}| = \mathrm{diag}(|s_x|, |s_y|, |s_z|)$. Define a diagonal matrix $\tilde{\boldsymbol{K}}_1$ as $\tilde{\boldsymbol{K}}_1 = \boldsymbol{K}_1 - |\tilde{\boldsymbol{D}}||\tilde{\boldsymbol{S}}|^{-1}$. According to Lemma 5.1, if $\tilde{\boldsymbol{K}}_1 > 0$, that is, $K_{1i} - |d_i|/|s_i| > 0$, then it holds that

$$\dot{V} \leq -\tilde{K}_{1\min}||\boldsymbol{s}||^2 - K_{2\min}||\boldsymbol{s}||^{\gamma+1} = -2\tilde{K}_{1\min}V - 2^{(\gamma+1)/2}K_{2\min}V^{(\gamma+1)/2} \tag{5.91}$$

where $\tilde{K}_{1\min}$ is the minimum eigenvalue of the matrix $\tilde{\boldsymbol{K}}_1$.

Then, according to Lemma 5.2, the region

$$|s_i| \le |d_i|/K_{1i} \le d_m/K_{1\min} \tag{5.92}$$

can be reached in finite time.

Similarly, from Eq. (5.90), it holds that

$$\dot{V} \le -\boldsymbol{s}^{\mathrm{T}}\boldsymbol{K}_1\boldsymbol{s} - \boldsymbol{s}^{\mathrm{T}}(\boldsymbol{K}_2 - |\tilde{\boldsymbol{D}}||\tilde{\boldsymbol{S}}^{\gamma}|^{-1})\mathrm{sig}^{\gamma}(\boldsymbol{s}) \tag{5.93}$$

where $|\tilde{\boldsymbol{S}}^{\gamma}| = \mathrm{diag}(|s_x|^{\gamma}, |s_y|^{\gamma}, |s_z|^{\gamma})$. Similarly, define a diagonal matrix $\tilde{\boldsymbol{K}}_2$ as $\tilde{\boldsymbol{K}}_2 = \boldsymbol{K}_2 - |\tilde{\boldsymbol{D}}||\tilde{\boldsymbol{S}}^{\gamma}|^{-1}$. Then, if $\tilde{\boldsymbol{K}}_2 > 0$, that is, $K_{2i} - |d_i|/|s_i|^{\gamma} > 0$, it holds that

$$\dot{V} \le -K_{1\min}||\boldsymbol{s}||^2 - \tilde{K}_{2\min}||\boldsymbol{s}||^{\gamma+1} = -2K_{1\min}V - 2^{(\gamma+1)/2}\tilde{K}_{2\min}V^{(\gamma+1)/2} \tag{5.94}$$

where $\tilde{K}_{2\min}$ is the minimum eigenvalue of the matrix $\tilde{\boldsymbol{K}}_2$.

Likewise, according to Lemma 5.2, the region

$$|s_i| \le (|d_i|/K_{2i})^{1/\gamma} \le (d_m/K_{2\min})^{1/\gamma} \tag{5.95}$$

can be reached in finite time. Thus, by evaluation of Eqs. (5.92) and (5.95), the finite-time convergent region of s_i can be derived as $|s_i| \le \varphi$, where the expression of φ has been given in Eq. (5.86).

After the region $|s_i| \le \varphi$ is reached, if $|e_i| \ge \delta$, the system dynamics is governed by

$$s_i = \dot{e}_i + \zeta_{1i}e_i + \zeta_{2i}e_i^{q/p}, \quad |s_i| \le \varphi \tag{5.96}$$

Rewrite Eq. (5.96) as

$$\dot{e}_i + (\zeta_{1i} - s_i e_i^{-1})e_i + \zeta_{2i}e_i^{q/p} = 0 \tag{5.97}$$

or

$$\dot{e}_i + \zeta_{1i}e_i + (\zeta_{2i} - s_i e_i^{-q/p})e_i^{q/p} = 0 \tag{5.98}$$

Then, according to Lemma 5.3, if $\zeta_{1i} - s_i e_i^{-1} > 0$ or $\zeta_{2i} - s_i e_i^{-q/p} > 0$ holds, the system errors will continue to converge. Thus, the terminal convergent region of e_i can be derived as

$$|e_i| \le \varphi/\zeta_{1i} \tag{5.99}$$

and

$$|e_i| \leq (\varphi/\zeta_{2i})^{p/q} \tag{5.100}$$

Furthermore, if $|e_i| < \delta$, it is easy to obtain that the state error e_i has been within the terminal convergent region $|e_i| \leq \varphi_{ei}$ as shown in Eq. (5.87). Then, using Eq. (5.96), the terminal convergent region of $\dot{e}_i$ can be derived as

$$|\dot{e}_i| \leq \varphi + \zeta_{1i}|e_i| + \zeta_{2i}|e_i|^{q/p} \leq \varphi + \zeta_{1i}\varphi_{ei} + \zeta_{2i}\varphi_{ei}^{q/p} \tag{5.101}$$

As can be seen, $\dot{e}_i$ will converge in finite time to the region $|\dot{e}_i| \leq \varphi_{\dot{e}i}$ as shown in Eq. (5.88). This completes the proof.

5.3.4 Numerical Simulation

5.3.4.1 Open-Loop Controller

Given that a Lorentz spacecraft is more efficient in LEOs, the chief's orbit is set as an elliptic LEO, and its initial orbit elements are listed in Table 5.2. The initial phase angle of the magnetic dipole is set as $\Omega_0 = -60°$, and the dipole tilt angle is $\alpha = 11.3°$.

It is assumed that the deputy starts from the origin of LVLH frame with zero relative velocity. Thus, the initial relative state vector (i.e., initial condition constraints) is $\boldsymbol{X}(t_0) = \mathbf{0}$. It is required to establish a formation configuration at the terminal time $t_f = T$, where T is the orbital period of the chief. As shown in Table 5.2, the chief is initially at the perigee of the orbit. Thus, at the terminal time $t_f = T$, the chief will return to the perigee, and the normalized orbital radius will be $\tilde{R}_C(t_f) = 1 - e_C$, where e_C is the eccentricity of the chief's orbit. At the terminal time $t_f = T$, the normalized radial velocity and orbital angular velocity are

$$\dot{\tilde{R}}_C(t_f) = 0, \quad \dot{\tilde{u}}_C(t_f) = (1 - e_C^2)^{-3/2}(1 + e_C)^2 \tag{5.102}$$

Furthermore, parts of the normalized terminal relative states are chosen as

Table 5.2 Initial orbit element of the chief spacecraft

Orbit element	Value
Semimajor axis (km)	7600
Eccentricity	0.1
Inclination (deg)	30
Right ascension of ascending node (deg)	50
Argument of perigee (deg)	0
True anomaly (deg)	0

Reprinted from Ref. [6], Copyright 2015, with permission from Elsevier

$$\begin{aligned} &\tilde{x}(t_f) = 0, && \tilde{y}(t_f) = 1.316 \times 10^{-4}, \quad \tilde{z}(t_f) = 0, \\ &\dot{\tilde{x}}(t_f) = 8.081 \times 10^{-5}, && \dot{\tilde{z}}(t_f) = 1.616 \times 10^{-4} \end{aligned} \tag{5.103}$$

which are corresponding to $y(t_f) = 10^3$ m, $\dot{x}(t_f) = 0.585$ m/s, and $\dot{z}(t_f) = 1.171$ m/s. By solving the energy-matching condition in Eq. (5.59) with the above states, a feasible solution to $\dot{\tilde{y}}(t_f)$ can be derived as $\dot{\tilde{y}}(t_f) = -2.551 \times 10^{-8}$. Now, the complete terminal relative states (i.e., final condition constraints) can be summarized as

$$\boldsymbol{X}(T) = \begin{bmatrix} 0 & 10^3 & 0 & 0.585 & -1.847 \times 10^{-4} & 1.171 \end{bmatrix}^{\mathrm{T}} \tag{5.104}$$

where the relative position and velocity are in unit m and m/s, respectively. The weight matrix is set as $\boldsymbol{Q} = \mathrm{diag}(1, 10^5, 10^5, 10^5)$ that the fuel consumption is given priority.

50 LG points are used in the GPM. The optimal hybrid control inputs derived via GPM are denoted by the discrete points in Fig. 5.9. To analyze the effect of the dipole tilt angle on the optimal solutions, a nontilted dipole model is also introduced to make comparisons. This nontilted model can be easily derived by setting $\alpha = 0$. In Fig. 5.9, the solid lines denote the Lagrangian interpolation results between the LG points. As can be seen, in both models, the specific charge necessary for

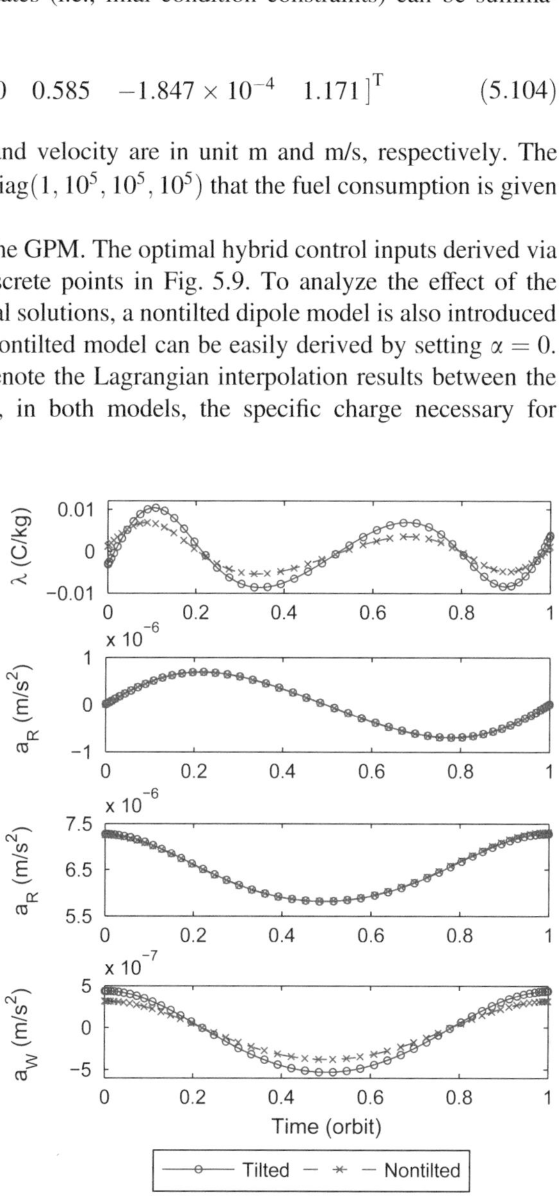

Fig. 5.9 Comparisons of time histories of control inputs. Reprinted from Ref. [6], Copyright 2015, with permission from Elsevier

formation establishment is smaller than the maximum 0.03 C/kg, and the required thruster-generated control acceleration is in the order of 10^{-6} m/s^2 or even smaller.

The corresponding Lorentz acceleration necessary for formation establishment is depicted in Fig. 5.10, from which obvious differences can be observed between these two models. The required Lorentz acceleration is generally in the order of 10^{-3} to 10^{-4} m/s^2, which is two or three orders of magnitude larger than that of the thruster-generated control acceleration. As aforementioned, J_2 perturbation is one of the most dominant perturbations in LEOs. For spacecraft within several kilometers in typical LEOs, the relative J_2-perturbation acceleration between spacecraft is in the order of about 10^{-5} to 10^{-6} m/s^2. Thus, as compared to the Lorentz acceleration and relative J_2-perturbation acceleration, the required thruster-generated control acceleration can be regarded as external perturbations in the closed-loop control system. As can be seen, with Lorentz force as auxiliary propulsion, the fuel consumptions can be greatly reduced, and the saved percent of velocity increments can be calculated via

$$\eta = 1 - \frac{\Delta V_C}{\Delta V_{\mathrm{UC}}} = 1 - \frac{\int_{t_0}^{t_f} \|\boldsymbol{a}_C\| \mathrm{d}t}{\int_{t_0}^{t_f} \|\boldsymbol{a}_L + \boldsymbol{a}_C\| \mathrm{d}t} \tag{5.105}$$

where ΔV_C and ΔV_{UC} refer to the velocity increment consumption necessary for a charged and uncharged spacecraft, respectively. For the tilted dipole model, if the deputy is uncharged that the total control acceleration is fully provided by thruster on board, the velocity increment consumption necessary for formation

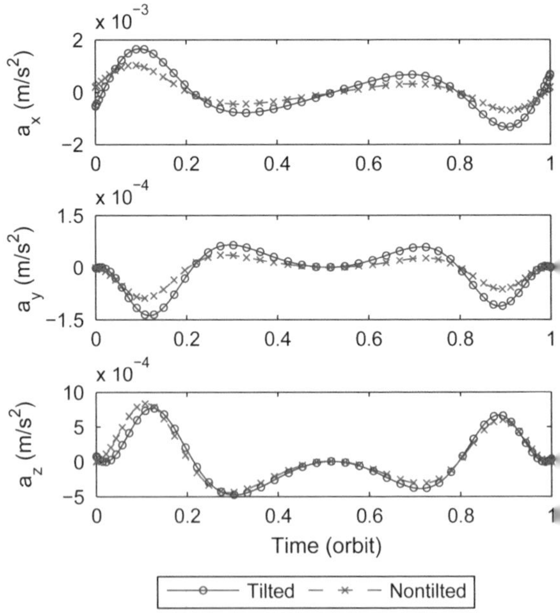

Fig. 5.10 Comparisons of time histories of Lorentz acceleration. Reprinted from Ref. [6], Copyright 2015, with permission from Elsevier

establishment is $\Delta V_{\mathrm{UC}} = 4.64$ m/s. However, if a charged Lorentz spacecraft is used, it decreases to $\Delta V_C = 0.04$ m/s, with about 99.1 % of the velocity increments saved. Thus, with Lorentz force as auxiliary propulsion, a nearly propellantless formation establishment is achieved in elliptic orbits. Likewise, for the nontilted dipole model, the velocity increments required by an uncharged and charged spacecraft are, respectively, $\Delta V_{\mathrm{UC}} = 3.07$ m/s and $\Delta V_C = 0.04$ m/s. Similarly, about 98.7 % of the velocity increments can be saved. Increasing the ratio of Q_u and Q_q can further decrease the velocity increment consumption.

Time histories of the relative position and velocity in both tilted and nontilted dipole model are compared in Figs. 5.11 and 5.12, respectively. Similarly, distinctive errors exist between these two models. The discrete points in both figures denote the results derived via GPM, but the solid and dashed lines are not the interpolated results any more. To verify the validity of the GPM's results, the interpolated optimal control inputs are substituted into the dynamical model in Eq. (5.1) to generate the real trajectory of relative position and velocity by numerical integration of the differential equations via fourth-order Runge–Kutta method. The numerical results are then denoted by the solid or dashed lines for the tilted or nontilted model. As can be seen, the numerical results are generally coincided with the GPM's results, thus verifying the validity of the GPM's results. The terminal relative position and velocity errors between the numerical and GPM's results are in the order of 10^{-2} m and 10^{-4} m/s, respectively, which are both within acceptable tolerances. Using more LG points and more precise integration methods could further decrease the errors. Moreover, if the optimal control inputs derived in

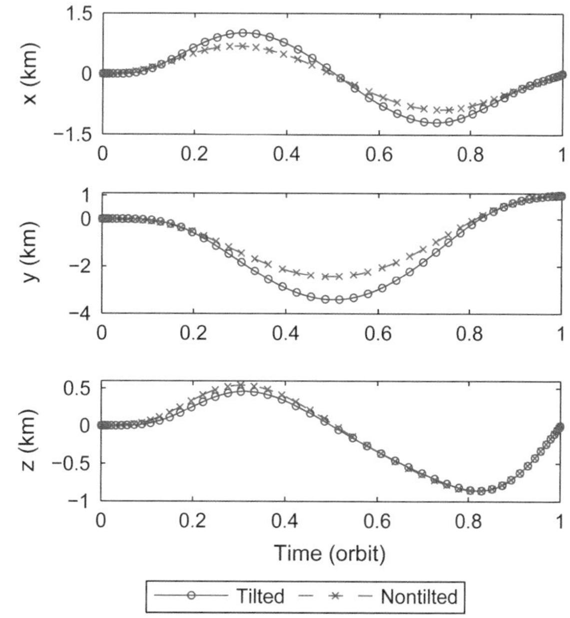

Fig. 5.11 Comparisons of time histories of relative position. Reprinted from Ref. [6], Copyright 2015, with permission from Elsevier

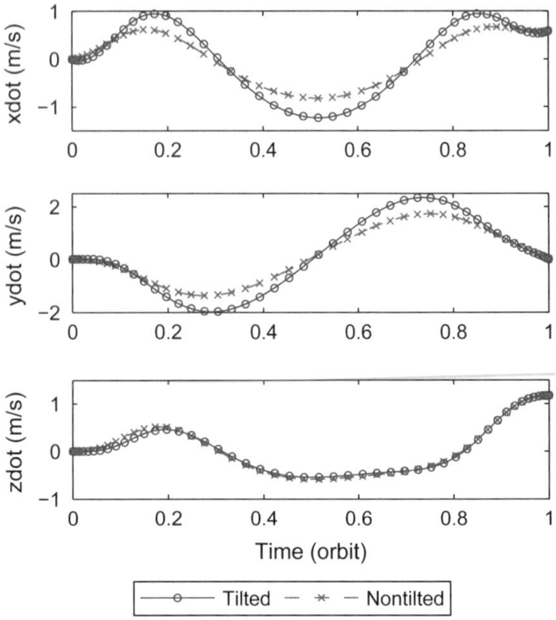

Fig. 5.12 Comparisons of time histories of relative velocity. Reprinted from Ref. [6], Copyright 2015, with permission from Elsevier

a nontilted dipole model are substituted into the tilted one, the resulting relative state error with respect to the desired one in Eq. (5.104) will be

$$\Delta \boldsymbol{X} = \begin{bmatrix} 22.9 & -142.5 & 50.5 & 0.02 & -0.04 & -0.47 \end{bmatrix}^{\mathrm{T}} \tag{5.106}$$

where the relative position error is in unit m, and the relative velocity error is in unit m/s. Such considerable error will definitely result in the failure of the desired formation establishment. Thus, in solving the optimal trajectories for Lorentz-augmented SFF in elliptic Earth orbits, the dipole tilt angle should be taken into account.

Figure 5.13 shows the relative transfer orbit and the desired formation configuration in the tilted dipole model. As can be seen, the deputy starts from the origin of the LVLH frame (i.e., the formation center), and establishes the desired formation configuration finally.

5.3.4.2 Closed-Loop Controller

The performance of the closed-loop controller is verified in this section. The initial orbit elements of the chief are also given in Table 5.2, and the initial phase angle of the dipole is also set as $\Omega_0 = -60°$. The initial relative position error is set as $\boldsymbol{e}(0) = \begin{bmatrix} 50 & -50 & 30 \end{bmatrix}^{\mathrm{T}}$ m, and the initial relative velocity error is set as zero, that

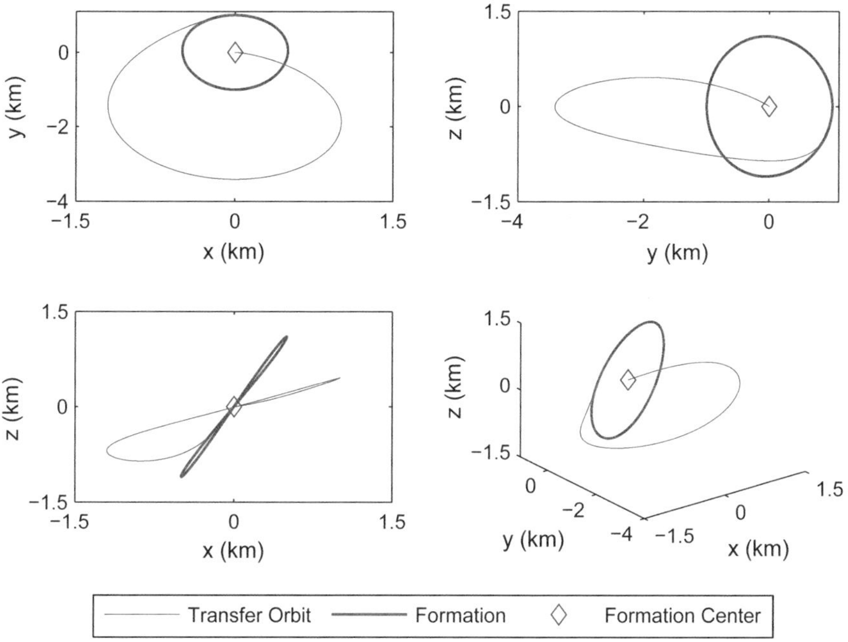

Fig. 5.13 Transfer orbit and formation trajectory of the tilted dipole model. Reprinted from Ref. [6], Copyright 2015, with permission from Elsevier

is, $\dot{\boldsymbol{e}}(0) = \mathbf{0}$. Given that J_2 perturbation is one of the most dominant perturbations in LEO, it is thus included as external disturbances. The other simulation parameters are summarized as follows: $p = 11$, $q = 9$, $\delta = 10^{-4}$, $\gamma = 3/5$, $d_m = 10^{-5}$ m/s^2, $\zeta_{1i} = 2 \times 10^{-3}$, $\zeta_{2i} = 10^{-3}$, $K_{1i} = 6 \times 10^{-3}$, and $K_{2i} = 3 \times 10^{-4}$ $(i = x, y, z)$. The desired relative states are those generated via GPM, as shown in the above section. Furthermore, to verify the faster convergent rate of FNTSMC than NTSMC, an NTSMC is also introduced to make comparisons. The NTSMC can easily be derived by setting $\zeta_{1i} = 0$.

Time histories of the relative position and velocity errors are shown in Figs. 5.14 and 5.15, respectively. Obviously, the convergent rate of FNTSMC is faster than that of NTSMC. For FNTSMC, the relative state errors converge to the neighborhood of zero after about $0.3T$, while it takes about one orbital period for the NTSMC to converge. For FNTSMC, the terminal control accuracy of the relative position and velocity error are, respectively, in the order of 10^{-2} m and 10^{-4} m/s. According to Theorem 5.1 and the above simulation parameters, it can be obtained that $\varphi = 1.67 \times 10^{-3}$. Numerical results indicate that the region $|s_i| \le \varphi$ is reached in a finite time of $t_r \approx 560$ s. Similarly, it can also be verified that the terminal convergent regions $|e_i| \le \varphi_{ei} = 0.84$ m and $|\dot{e}_i| \le \varphi_{\dot{e}i} = 4.20 \times 10^{-3}$ m/s are also reached in finite time. In this way, the finite-time convergence of the closed-loop FNTSMC has been proved. Furthermore, by increasing $K_{j\min}$ $(j = 1, 2)$, the system

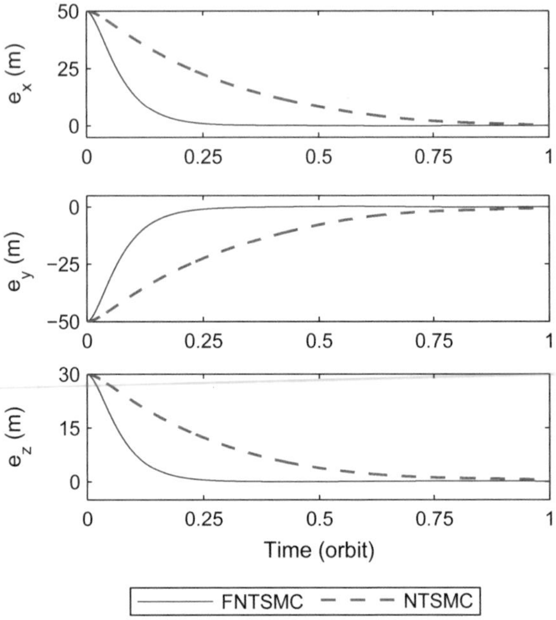

Fig. 5.14 Time histories of relative position errors. Reprinted from Ref. [6], Copyright 2015, with permission from Elsevier

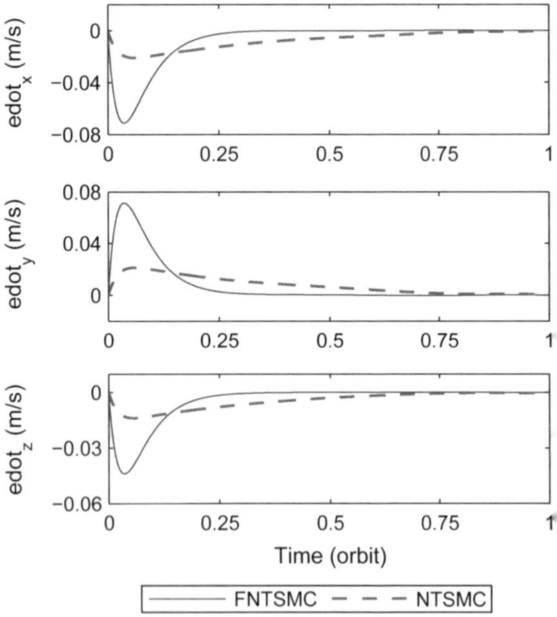

Fig. 5.15 Time histories of relative velocity errors. Reprinted from Ref. [6], Copyright 2015, with permission from Elsevier

trajectory will be closer to the sliding surface $\boldsymbol{s} = \mathbf{0}$. Meanwhile, increasing ζ_{ji} $(i = x, y, z;\ j = 1, 2)$ and decreasing δ can further increase the terminal control accuracy.

Time histories of the control inputs are compared in Fig. 5.16. As can be seen, for both controllers, the control inputs are continuous and no chattering is observed. For a charged Lorentz spacecraft, the velocity increment necessary for formation establishment consumed by the FNTSMC and NTSMC are $\Delta V_F = 0.27$ m/s and $\Delta V_N = 0.24$ m/s, respectively. Obviously, despite faster convergent rate, a bit more velocity increment is consumed by the FNTSMC than NTSMC. Furthermore, for an uncharged spacecraft, the corresponding velocity increment consumption will increase to $\Delta V_F = 4.71$ m/s and $\Delta V_N = 4.58$ m/s, respectively. Thus, with Lorentz force as auxiliary propulsion, about 94 % of the velocity increment can be saved in this example. Aforementioned simulation results verify the validity of the proposed FTNSMC for Lorentz-augmented formation establishment. Notably, this closed-loop controller can also be applied to other Lorentz-augmented relative orbital control problems, such as rendezvous and formation reconfiguration.

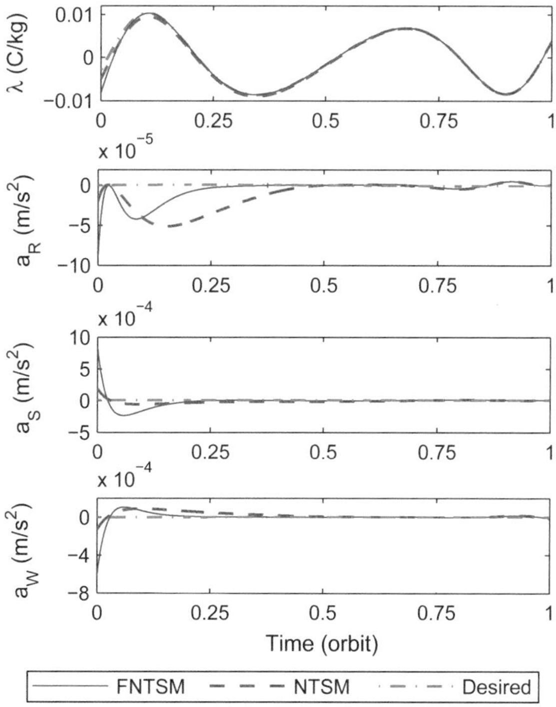

Fig. 5.16 Time histories of control inputs. Reprinted from Ref. [6], Copyright 2015, with permission from Elsevier

5.4 Conclusions

This chapter analyzes the dynamics and control of Lorentz-augmented SFF in both circular and elliptic orbits. An open-loop controller is first derived via the pseudospectral method, and a closed-loop fast nonsingular terminal sliding mode controller is designed to ensure finite-time tracking of the optimal open-loop trajectory. For SFF in circular orbits, both fixed and free final conditions are used to make comparisons. Simulation results indicate that using free final conditions, the control energy can be saved. Notably, the closed-loop controller proposed for the elliptic reference orbits could also be applicable to circular reference orbits. Also, simulation results verify the finite-time convergence and fast convergent rate of the closed-loop controller.

References

1. Pollock GE, Gangestad JW, Longuski JM (2011) Analytical solutions for the relative motion of spacecraft subject to Lorentz-force perturbations. Acta Astronaut 68:204–217
2. Clohessy W, Wiltshire R (1960) Terminal guidance system for satellite rendezvous. J Aerosp Eng 27:653–658
3. Sabol C, Burns R, McLaughlin CA (2001) Satellite formation flying design and evolution. J Spacecr Rockets 38:270–278
4. Benson DA, Huntington GT, Thorvaldsen TP et al (2006) Direct trajectory optimization and costate estimation via an orthogonal collocation method. J Guid Control Dyn 29:1435–1440
5. Huang X, Yan Y, Zhou Y (2014) Optimal spacecraft formation establishment and reconfiguration propelled by the geomagnetic Lorentz force. Adv Space Res 54:2318–2335
6. Huang X, Yan Y, Zhou Y (2015) Optimal Lorentz-augmented spacecraft formation flying in elliptic orbits. Acta Astronaut 111:37–47
7. Gurfil P (2005) Relative motion between elliptic orbits: generalized boundedness conditions and optimal formationkeeping. J Guid Control Dyn 28:761–767
8. Zou A, Kumar D, Hou Z et al (2011) Finite-time attitude tracking control for spacecraft using terminal sliding mode and Chebyshev neural network. IEEE Trans Syst Man Cybern B Cybern 41:950–963
9. Yu S, Yu X, Shirinzadeh B et al (2005) Continuous finite-time control for robotic manipulators with terminal sliding mode. Automatica 41:1957–1964
10. Yu X, Man Z (2002) Fast terminal sliding-mode control design for nonlinear dynamical systems. IEEE Trans Circuits Syst I Fundam Theory Appl 49:261–264

Chapter 6 Dynamics and Control of Lorentz-Augmented Spacecraft Rendezvous

This chapter investigates the dynamics and control problem of Lorentz-augmented spacecraft rendezvous. Previous propellantless rendezvous strategy is developed based on a linearized dynamical model of Lorentz-augmented relative orbital motion. Due to the characteristic of the linearized dynamics, several constraints are imposed on the initial relative position and maneuvering time in this strategy. To get rid of these constraints, nonlinear optimal control schemes for Lorentz spacecraft rendezvous are designed via pseudospectral method in this chapter, based on the nonlinear model of Lorentz-augmented relative motion. In designing the novel rendezvous strategy, two cases with both fixed and free final time are discussed. Also, the effect of J_2 perturbation on optimal rendezvous trajectory is analyzed.

6.1 Problem Formulation and Dynamical Model

6.1.1 Problem Formulation

Previous rendezvous strategy is developed based on a linearized model. In this strategy, it is required that the chaser and the target fly in the same circular equatorial orbit, and that the target initially leads or lags the chaser in the in-track direction by several kilometers with no relative velocity. Furthermore, the maneuvering time should equal an integer number of the reference orbit periods and an integer number of the rotation periods of Earth. Then, maintaining a constant specific charge could achieve rendezvous with the given maneuvering time. As can be seen, strict restrictions are imposed on the chief's orbit, initial relative state, and maneuvering time. Therefore, to get rid of these restrictions, it is necessary to design novel rendezvous strategies. The problems discussed in this chapter are now summarized as follows.

© Springer Science+Business Media Singapore 2017

Y. Yan et al., *Dynamics and Control of Lorentz-Augmented Spacecraft Relative Motion*, DOI 10.1007/978-981-10-2603-4_6

1. Design optimal open-loop rendezvous schemes in both circular and elliptic orbits by using the pseudospectral method. Both fixed and free final time are considered to make comparisons;
2. Design closed-loop controllers to ensure optimal trajectory tracking in the presence of external disturbances;
3. Analyze the effect of J_2 perturbation on the optimal open-loop trajectory.

6.1.2 Dynamical Model

According to Sect. 2.2.1.1 and based on the same definitions of coordinate frames, the dynamic equations of Lorentz-augmented relative motion can be rewritten in the target's LVLH frame as

$$\begin{aligned} \ddot{x} &= 2\dot{u}_T\dot{y} + \dot{u}_T^2 x + \ddot{u}_T y + n_T^2 R_T - n_L^2(R_T + x) + a_x + a_R \\ \ddot{y} &= -2\dot{u}_T\dot{x} + \dot{u}_T^2 y - \ddot{y}_T x - n_L^2 y + a_y + a_S \\ \ddot{z} &= -n_L^2 z + a_z + a_W \end{aligned} \tag{6.1}$$

where $\boldsymbol{\rho} = \begin{bmatrix} x & y & z \end{bmatrix}^{\mathrm{T}}$ is the relative position vector between the chaser and the target. $n_L = \sqrt{\mu/R_L^3}$ and $n_T = \sqrt{\mu/R_T^3}$, with $R_L = \left[(R_T + x)^2 + y^2 + z^2\right]^{1/2}$ and R_T being the orbital radius of the chaser and target spacecraft, respectively. Notably, the chaser is a charged Lorentz spacecraft, and the target is uncharged. u_T is the argument of latitude of the target, and $\dot{u}_T$ and $\ddot{u}_T$ are thus the orbital angular velocity and acceleration of the target, respectively. Also, $\boldsymbol{a}_L = \begin{bmatrix} a_x & a_y & a_z \end{bmatrix}^{\mathrm{T}}$ and $\boldsymbol{a}_C = \begin{bmatrix} a_R & a_S & a_W \end{bmatrix}^{\mathrm{T}}$ are the Lorentz acceleration and thruster-generated control acceleration acting on the chaser, respectively.

The Lorentz acceleration can be expressed in the LVLH frame as

$$\boldsymbol{a}_L = (q_L/m_L)\boldsymbol{V}_{\mathrm{rel}} \times \boldsymbol{B} \tag{6.2}$$

where $\lambda = q_L/m_L$ is the specific charge of Lorentz spacecraft. $\boldsymbol{B}$ is the magnetic field local at the Lorentz spacecraft, given by [1]

$$\boldsymbol{B} = \begin{bmatrix} B_x & B_y & B_z \end{bmatrix}^{\mathrm{T}} = (B_0/R_L^3)\left[3(\hat{\boldsymbol{n}} \cdot \hat{\boldsymbol{R}}_L)\hat{\boldsymbol{R}}_L - \hat{\boldsymbol{n}}\right] \tag{6.3}$$

where $\hat{\boldsymbol{R}}_L = (1/R_L)\begin{bmatrix} R_T + x & y & z \end{bmatrix}^{\mathrm{T}}$ is the unit orbital radius vector of the Lorentz spacecraft, and $\hat{\boldsymbol{n}}$ is the unit magnetic dipole momentum vector, given by

$$\hat{\boldsymbol{n}} = \begin{bmatrix} n_x \\ n_y \\ n_z \end{bmatrix} = \begin{bmatrix} -(\cos\beta\cos u_T + \sin\beta\cos i_T\sin u_T)\sin\alpha - \sin i_T\sin u_T\cos\alpha \\ (\cos\beta\sin u_T - \sin\beta\cos i_T\cos u_T)\sin\alpha - \sin i_T\cos u_T\cos\alpha \\ \sin\beta\sin i_T\sin\alpha - \cos i_T\cos\alpha \end{bmatrix} \tag{6.4}$$

where i_T is the orbital inclination of the target, and α is dipole tilt angle. The angle β is defined as $\beta = \Omega_M - \Omega_T$. Ω_T is the right ascension of the ascending node of the target, and $\Omega_M = \omega_E t + \Omega_0$ is the inertial phase angle of the magnetic dipole. Also, ω_E is the rotation rate of Earth, t is the time, and Ω_0 is the initial phase angle of the dipole.

Given that the Earth's magnetic field is corotating with Earth, the velocity of the Lorentz spacecraft relative to the magnetic field is thus given by

$$\boldsymbol{V}_{\text{rel}} = \begin{bmatrix} V_x \\ V_y \\ V_z \end{bmatrix} = \begin{bmatrix} \dot{R}_T + \dot{x} - y(\dot{u}_T - \omega_E\cos i_T) - z\omega_E\sin i_T\cos u_T \\ \dot{y} + (R_T + x)(\dot{u}_T - \omega_E\cos i_T) + z\omega_E\sin i_T\sin u_T \\ \dot{z} + (R_T + x)\omega_E\sin i_T\cos u_T - y\omega_E\sin i_T\sin u_T \end{bmatrix} \tag{6.5}$$

Substitution of the expression of $\boldsymbol{a}_L$ into Eq. (6.1) yields the dynamical model of Lorentz-augmented spacecraft rendezvous.

6.2 Case of Two-Body Reference Orbit

A two-body reference orbit is assumed in this section, whether a circular or an elliptic one. For an elliptic Keplerian orbit, the orbital motion of the target is governed by [2]

$$\begin{aligned} \ddot{R}_T &= R_T\dot{u}_T^2 - \mu/R_T^2 \\ R_T\ddot{u}_T &= -2\dot{R}_T\dot{u}_T \end{aligned} \tag{6.6}$$

6.2.1 *Open-Loop Control*

Similar to the application of Lorentz-augmented SFF, the optimal control problem of Lorentz-augmented rendezvous is first formulated as a general TOP. For the case with two-body reference orbit, define the specific charge of Lorentz spacecraft (i.e., $\lambda = q_L/m_L$) as the single control input, that is, $\boldsymbol{a}_C = 0$. Find the optimal profile of λ and the charging time (i.e., the maneuvering time t_f) that minimizes the following objective function:

$$J = \int_{t_0}^{t_f} \lambda^2 \mathrm{d}t \tag{6.7}$$

The dynamic constraints are given in Eqs. (6.1) and (6.6). Other constraints, including the boundary and path constraints, are summarized as follows.

The boundary constraints consist of the initial and final condition constraints. The initial condition constraints are just the initial relative states between the chaser and the target, given by

$$\begin{aligned} x(t_0) &= x_0,\ y(t_0) = y_0, z(t_0) = z_0 \\ \dot{x}(t_0) &= \dot{x}_0,\ \dot{y}(t_0) = \dot{y}_0, \dot{z}(t_0) = \dot{z}_0 \end{aligned} \tag{6.8}$$

To achieve rendezvous at the final time t_f, it is required that the two spacecraft reach the same position with zero relative velocity. Thus, the terminal condition constraints are just the terminal relative states between the chaser and the target, given by

$$\begin{aligned} x(t_f) &= 0, y(t_f) = 0, z(t_f) = 0 \\ \dot{x}(t_f) &= 0, \dot{y}(t_f) = 0, \dot{z}(t_f) = 0 \end{aligned} \tag{6.9}$$

Similarly, given that the near-term feasible maximum of the specific charge is about 0.03 C/kg [3], a path constraint imposed on the control input is proposed as

$$-0.03 \leq \lambda \leq 0.03 \tag{6.10}$$

By formulating the Lorentz-augmented rendezvous problem as the aforementioned TOP, the optimal open-loop control trajectory can then be solved by the GPM. Detailed description of GPM is referred to Sect. 5.2.3.

6.2.2 Closed-Loop Control

As stated in Sect. 4.3.2 or Sect. 5.3.3, the closed-loop controller for Lorentz-augmented spacecraft hovering or formation flying could also be applied to other Lorentz-augmented relative orbital control problems, such as rendezvous. Therefore, detailed descriptions of the closed-loop control scheme for Lorentz-augmented rendezvous are not presented here for brevity, and the reader is referred to Sect. 4.3.2 or Sect. 5.3.3 for details.

6.2.3 *Numerical Simulation*

The target is assumed to be flying in a near-circular LEO with an orbital period of 1.6 h, and the initial orbit elements of the target are listed in Table 6.1. The initial phase angle of the magnetic dipole is set as $\Omega_0 = 40°$. Two cases with fixed and free final time are considered. In the first case, the final time is set as a predetermined fixed point. However, in the second case, the final time is not determined in advance, and is regarded as an optimization variable. In both cases, 50 LG points are used in the GPM, and the initial relative states (i.e., the initial condition constraints) are set as

$$\begin{aligned} x_0 &= -250\text{ m},\ y_0 = -1000\text{ m},\ z_0 = -200\text{ m} \\ \dot{x}_0 &= -0.020\text{ m/s},\ \dot{y}_0 = 0.545\text{ m/s},\ \dot{z}_0 = 0.090\text{ m/s} \end{aligned} \tag{6.11}$$

6.2.3.1 Case with Fixed Final Time

The final time is fixed as $t_f = 5000$ s. Figure 6.1 shows the optimal trajectory of the specific charge necessary for rendezvous, from which it is obvious that the specific charge is on the order of 10^{-3} C/kg, smaller than the maximum of 0.03 C/kg. The discrete points in Fig. 6.1 denote the results generated via GPM, and the solid line denotes the Lagrangian interpolation results between the LG points. The corresponding control energy consumption is $J = 1.15 \times 10^{-2}$.

Time histories of the relative position and velocity are shown in Figs. 6.2 and 6.3, respectively. Also, the discrete points denote the results generated via GPM. To verify the validity of the GPM's results, the interpolated trajectory of specific charge is substituted into the dynamical model [i.e., Eq. (6.1)] to derive the real trajectory of relative position and velocity by numerically integrating the nonlinear equations via the 4th order Runge–Kutta method. The resulting numerical trajectories of relative position and velocity are denoted by the solid lines in Figs. 6.2 and 6.3, respectively. As can be seen, the numerical results are nearly coincided with the GPM's results, and rendezvous is achieved at the final time. Specifically, the relative position errors between the numerical and GPM's results are, respectively,

Table 6.1 Initial orbit elements of the target spacecraft

Orbit element	Value
Semi-major axis (km)	6945.034
Eccentricity	0.0002
Inclination (deg)	30
Right ascension of ascending node (deg)	30
Argument of perigee (deg)	10
True anomaly (deg)	10

Reprinted from Ref. [3], with kind permission from Springer Science+Business Media

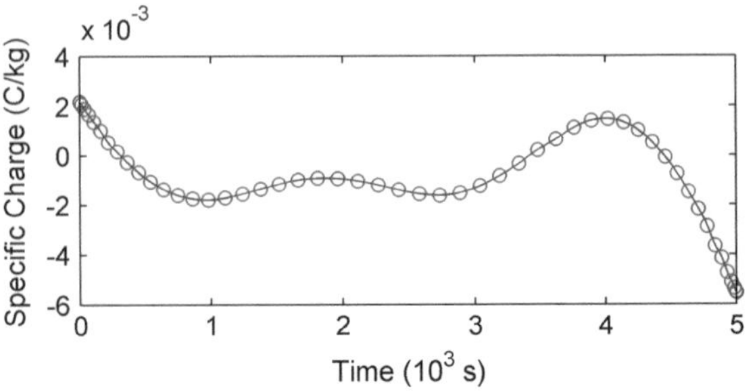

Fig. 6.1 Time histories of specific charge ($t_f = 5000$ s). Reprinted from Ref. [3], with kind permission from Springer Science+Business Media

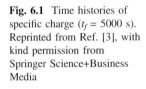

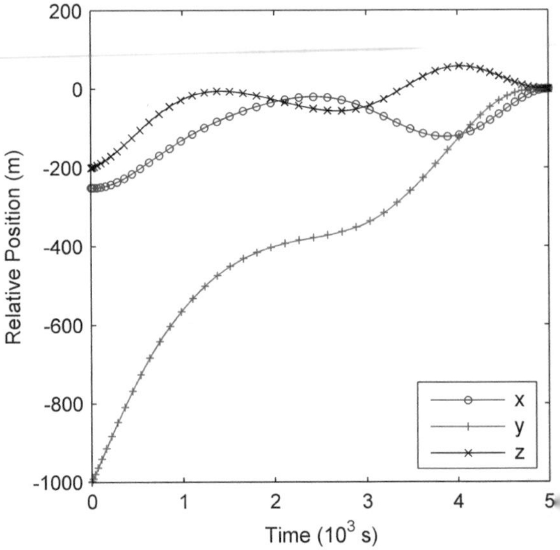

Fig. 6.2 Time histories of relative position ($t_f = 5000$ s). Reprinted from Ref. [3], with kind permission from Springer Science+Business Media

0.52, 0.43 and 0.008 m, in the radial, in-track, and normal directions. Furthermore, the relative velocity errors between these two methods are 1.15×10^{-4} m/s, 1.10×10^{-3} m/s and 5.20×10^{-6} m/s in each of the directions. All errors are within acceptable tolerances, indicating the validity of the GPM's results. These errors arise from the interpolation and integration methods, which can be further reduced by choosing more LG points and more precise integration methods.

6.2.3.2 Case with Free Final Time

Differently, the final time in this case is not fixed, but is required to satisfy $t_f < 10^4$ s. The resulting optimal trajectory of the specific charge is shown in Fig. 6.4. Also, the discrete points denote the GPM's results, and the solid line refers to the Lagrangian interpolated result between the LG points. As can be seen, the necessary specific charge for this case is generally on the order of 10^{-4} C/kg, one

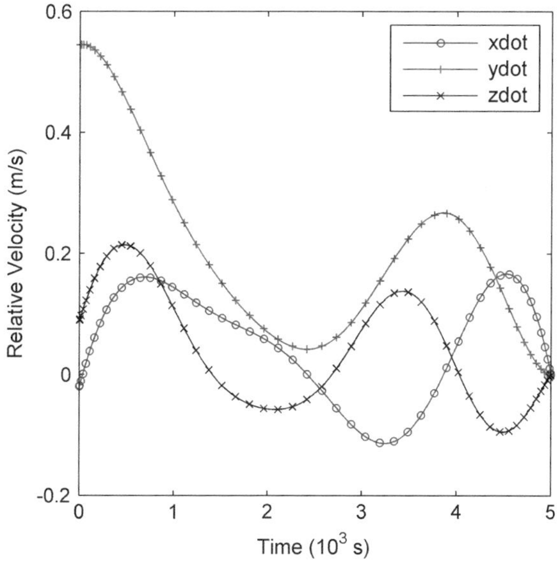

Fig. 6.3 Time histories of relative velocity ($t_f = 5000$ s). Reprinted from Ref. [3], with kind permission from Springer Science+Business Media

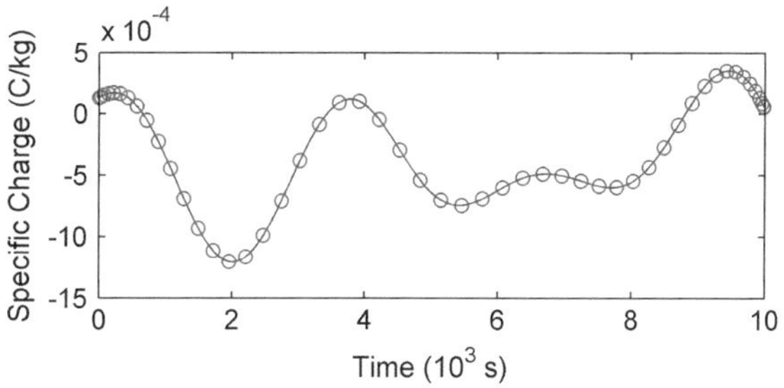

Fig. 6.4 Time histories of specific charge ($t_f < 10^4$ s). Reprinted from Ref. [3], with kind permission from Springer Science+Business Media

order of magnitude smaller than that for the case with fixed final time. Rendezvous is achieved at the final time $t_f = 9984.2\,\mathrm{s}$, with a corresponding energy consumption of $J = 3.20 \times 10^{-3}$. Compared to the former case, the charging time is nearly doubled, but 72.17 % of the control energy is saved.

Time histories of the relative position and velocity for this case are depicted in Figs. 6.5 and 6.6, respectively. Likewise, the GPM's results are denoted by the discrete points. Then, to verify the validity of the GPM's results, the interpolated trajectory of the specific charge is also substituted into the dynamical model to generate numerical trajectory of real relative position and velocity, as shown by the solid lines in Figs. 6.5 and 6.6, respectively. Similarly, the numerical results are nearly coincided with the GPM's results, and the terminal relative errors are also all within acceptable tolerances. By using these two examples, the feasibility and validity of GPM in solving the Lorentz-augmented rendezvous problem with either fixed or free final time have been substantiated.

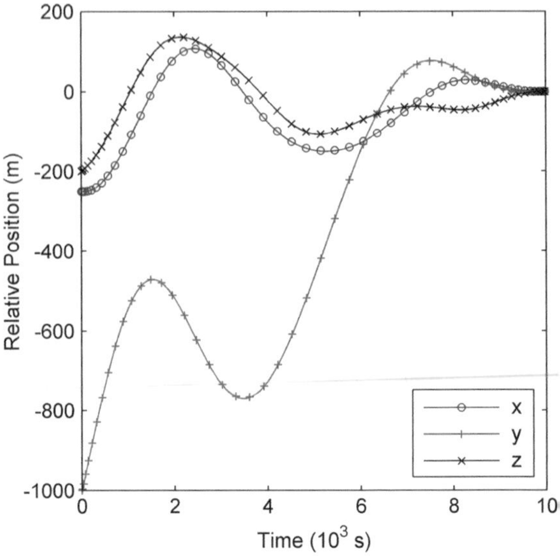

Fig. 6.5 Time histories of relative position ($t_f < 10^4$ s). Reprinted from Ref. [3], with kind permission from Springer Science+Business Media

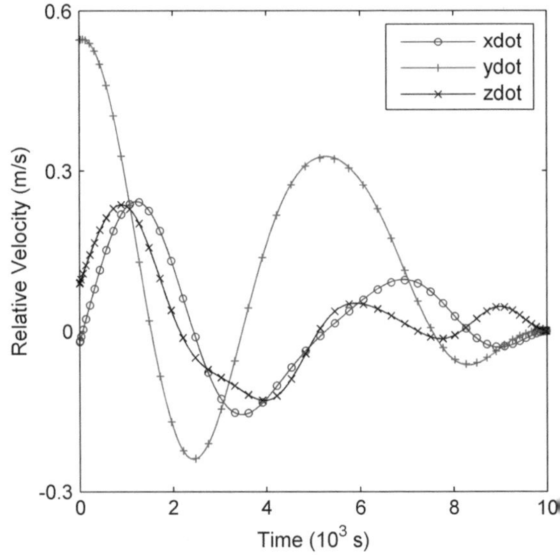

Fig. 6.6 Time histories of relative velocity ($t_f < 10^4$ s). Reprinted from Ref. [3], with kind permission from Springer Science+Business Media

6.3 Case of J_2-Perturbed Reference Orbit

6.3.1 Dynamical Model

According to the analysis in Sect. 2.2.3, the dynamical model of J_2-perturbed Lorentz-augmented spacecraft rendezvous can be rewritten as follows [4]:

$$\begin{cases} \ddot{x} = 2\dot{y}\omega_z - x(\eta_L^2 - \omega_z^2) + y\varepsilon_z - z\omega_x\omega_z - (\xi_L - \xi)\sin i_T \sin u_T \\ \quad -R_T(\eta_L^2 - \eta^2) + a_x + a_R \\ \ddot{y} = -2\dot{x}\omega_z + 2\dot{z}\omega_x - x\varepsilon_z - y(\eta_L^2 - \omega_z^2 - \omega_x^2) + z\varepsilon_x \\ \quad -(\xi_L - \xi)\sin i_T \cos u_T + a_y + a_S \\ \ddot{z} = -2\dot{y}\omega_x - x\omega_x\omega_z - y\varepsilon_x - z(\eta_L^2 - \omega_x^2) - (\xi_L - \xi)\cos i_T \\ \quad + a_z + a_W \end{cases} \tag{6.12}$$

where the detailed definitions of the variables are referred to Sect. 2.2.3 or Sect. 4.4.1. For the purpose of brevity, the definitions are not listed here again.

In the presence of J_2 perturbation, the six orbital parameters of the target are governed by [5]

$$\begin{cases} \dot{R}_T = V_r \\ \dot{V}_r = -\mu/R_T^2 + h_T^2/R_T^3 - k_J(1 - 3\sin^2 i_T \sin^2 u_T)/R_T^4 \\ \dot{h}_T = -(k_J \sin^2 i_T \sin 2u_T)/R_T^3 \\ \dot{i}_T = -(k_J \sin 2i_T \sin 2u_T)/(2h_T R_T^3) \\ \dot{u}_T = h_T/R_T^2 + (2k_J \cos^2 i_T \sin^2 u_T)/(h_T R_T^3) \\ \dot{\Omega}_T = -(2k_J \cos i_T \sin^2 u_T)/(h_T R_T^3) \end{cases} \tag{6.13}$$

Similarly, the definitions of the above variables have been elaborated in Sect. 2.2.3 or Sect. 4.4.1.

6.3.2 *Open-Loop Control*

As to the case with a two-body reference orbit, the optimal control problem of J_2-perturbed Lorentz-augmented spacecraft rendezvous is formulated as a TOP. The energy-optimal objective function is also chosen as Eq. (6.7). Furthermore, the dynamic constraints are given in Eqs. (6.12) and (6.13), the boundary constraints are given in Eqs. (6.8) and (6.9), and the path constraint is Eq. (6.10).

Likewise, by solving the resulting TOP via GPM, the energy-optimal open-loop control strategies in the J_2-perturbed environment can be derived.

6.3.3 *Numerical Simulations*

The target is assumed to be flying in a near-circular LEO, and its initial orbital elements are listed in Table 6.2. The initial phase angle of the magnetic dipole is set as $\Omega_0 = -60°$. It is required that the rendezvous is achieved at the finial time predetermined as $t_f = 6000$ s.

Table 6.2 Initial orbital elements of the target spacecraft

Orbit element	Value
Semi-major axis (km)	6945.034
Eccentricity	0.0002
Inclination (deg)	30
Right ascension of ascending node (deg)	50
Argument of perigee (deg)	20
True anomaly (deg)	−20

Reprinted from Ref. [4], Copyright 2015, with permission from SAGE

Also, it is assumed that the target initially lags off the target by 1 km in the in-track direction with zero relative velocity. Thus, the initial boundary constraints (i.e., initial relative states) are given by

$$\begin{aligned} x_0 &= 0 \text{ m},\ y_0 = -1000 \text{ m},\ z_0 = 0 \text{ m} \\ \dot{x}_0 &= 0 \text{ m/s},\ \dot{y}_0 = 0 \text{ m/s},\ \dot{z}_0 = 0 \text{ m/s} \end{aligned} \tag{6.14}$$

60 LG points are used in the GPM to solve the optimal trajectory. Notably, to investigate the effect of J_2 perturbation on the optimal trajectory, the optimal results derived in an unperturbed model (i.e., the two-body one) are also introduced to make comparisons. For both J_2-perturbed and unperturbed models, the energy-optimal trajectories of specific charge necessary for rendezvous are compared in Fig. 6.7. Also, the discrete points denote the results generated via GPM,

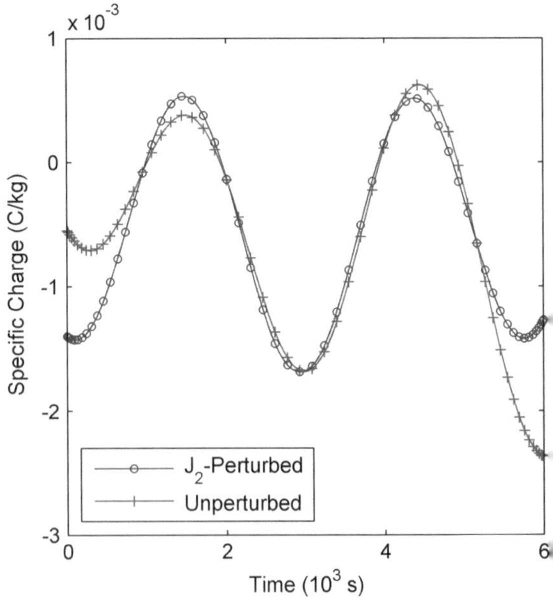

Fig. 6.7 Comparisons of the optimal trajectories of specific charge. Reprinted from Ref. [4], Copyright 2015, with permission from SAGE

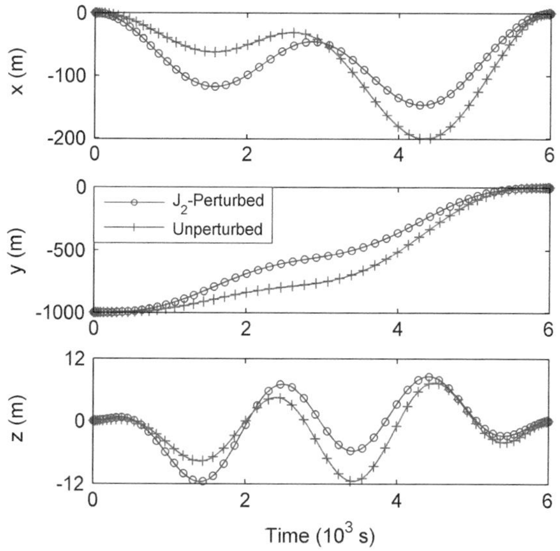

Fig. 6.8 Comparisons of the optimal trajectories of relative position. Reprinted from Ref. [4], Copyright 2015, with permission from SAGE

and the solid lines denote the Lagrangian interpolation results between the LG points. As can be seen, for both models, the required specific charge is on the order of 10^{-3} C/kg, within the maximum of 0.03 C/kg.

Time histories of the relative position and velocity are then shown in Figs. 6.8 and 6.9, respectively. Likewise, the discrete points denote the GPM's results. To verify the validity of the GPM's results, the interpolated trajectories of the optimal specific charge are then substituted into the corresponding dynamical model to derive the real trajectory of relative position and velocity via numerical integrations of the dynamic equations by using 4th order Runge–Kutta method. The numerical results are then denoted by the solid lines in both figures. As can be seen, the numerical results are generally coincided with the GPM's results, verifying the validity of GPM.

Furthermore, as shown in Figs. 6.8 and 6.9, rendezvous is achieved at the final time for both models, but distinctive errors exist between the rendezvous trajectories in these two models. If the optimal trajectory of specific charge derived in an unperturbed model is inputted in the J_2-perturbed model, then the resulting terminal relative position error will be 0.44, 6.54 and −0.96 m in each of the radial, in-track, and normal directions, indicating of the failure of the rendezvous mission. Thus, when designing Lorentz-augmented rendezvous strategies in typical LEOs, the J_2 perturbation should be taken into account.

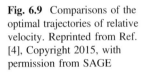

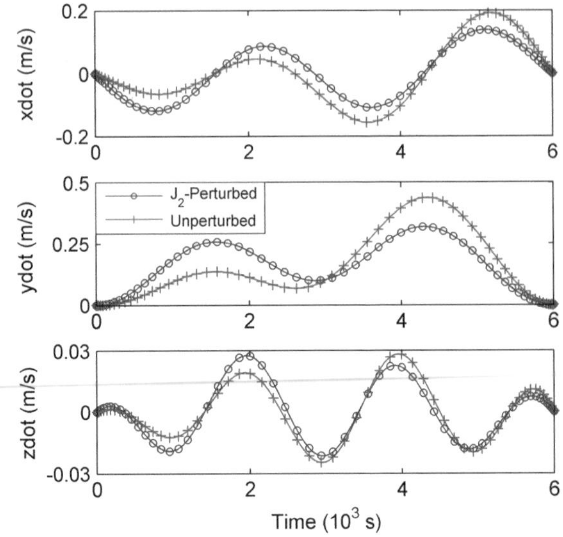

Fig. 6.9 Comparisons of the optimal trajectories of relative velocity. Reprinted from Ref. [4], Copyright 2015, with permission from SAGE

6.4 Conclusions

This chapter investigates the feasibility of using the geomagnetic Lorentz force for spacecraft rendezvous in circular or elliptic orbits.

Based on the dynamical model of Lorentz-augmented spacecraft relative motion, the optimal control problem of Lorentz-augmented rendezvous is firstly formulated as a constrained TOP. The resulting TOP is then solved by GPM to derive the optimal trajectories. Meanwhile, given that J_2 perturbation is one of the most dominant perturbations in LEO, the optimal trajectories in J_2-perturbed environment are also introduced to make comparisons with the unperturbed ones. The comparison results indicate that J_2 perturbation should be taken into account when designing the rendezvous strategies in typical LEOs.

References

1. Pollock GE, Gangestad JW, Longuski JM (2011) Analytical solutions for the relative motion of spacecraft subject to Lorentz-force perturbations. Acta Astronaut 68:204–217
2. Tsujii S, Bando M, Yamakawa H (2013) Spacecraft formation flying dynamics and control using the geomagnetic Lorentz force. J Guid Control Dyn 36:136–148
3. Huang X, Yan Y, Zhou Y et al (2015) Pseudospectral method for optimal propellantless rendezvous using geomagnetic Lorentz force. Appl Math Mech Engl Ed 36:609–618
4. Huang X, Yan Y, Zhou Y et al (2015) Nonlinear relative dynamics of Lorentz spacecraft about J_2-perturbed orbit. Proc Inst Mech Eng Part G J Aerosp Eng 229:467–478. doi:10.1177/0954410014537231
5. Xu G, Wang D (2008) Nonlinear dynamic equations of relative motion around an oblate Earth. J Guid Control Dyn 31:1521–1524

Appendix

A.1 Coefficients in Eqs. (2.56) and (2.57)

The coefficients in Eqs. (2.56) and (2.57) are summarized below.

$$\eta = \frac{q_L}{m_L}\frac{B_0}{R_T^3} \tag{A.1}$$

$$\begin{aligned}\kappa_{t,0}^x = &-\frac{1}{2}\eta\cos\alpha\sin i_T\left[(z_0 + k_{z1})n_T\sin\varphi_3 + (\dot{z}_0 + k_{z2})\cos\varphi_3 - \frac{c_3}{4n_T}\right]\\ &+ R_T\eta\cos\alpha\left[\cos i_T(n_T - \omega_E\cos i_T) - \frac{1}{2}\sin^2 i_T\omega_E\right]\\ &+ \frac{1}{4}\eta\sin\alpha\left[\frac{c_1(n_T+\omega_E)}{\omega_E(\omega_E+2n_T)}(1-\cos i_T) + \frac{c_2(n_T-\omega_E)}{\omega_E(\omega_E-2n_T)}(1+\cos i_T)\right]\end{aligned} \tag{A.2}$$

$$\kappa_{t,0}^y = \eta\cos\alpha\sin i_T[(z_0 + k_{z1})n_T\cos\varphi_3 - (\dot{z}_0 + k_{z2})\sin\varphi_3] \tag{A.3}$$

$$\kappa_{t,1}^y = -\eta\frac{1}{2}c_3\cos\alpha\sin i_T \tag{A.4}$$

$$\begin{aligned}A_{m,i}^x &= a_{m,i}^x n_T + b_{m,i}^x\omega_E, \quad A_{m,i}^y = a_{m,i}^y n_T + b_{m,i}^y\omega_E\\ B_{m,j}^x &= c_{m,j}^x n_T + d_{m,j}^x\omega_E, \quad B_{m,j}^y = c_{m,j}^y n_T + d_{m,j}^y\omega_E\end{aligned} \tag{A.5}$$

with

$C_{1,1}^x = \frac{1}{2}(z_0 + k_{z1})\eta\cos\alpha\sin i_T n_T$ $\qquad a_{1,1}^x = 2 \quad b_{1,1}^x = 0 \quad \theta_{1,1}^x = \varphi_3$

$C_{1,2}^x = \frac{1}{4}(\dot{z}_0 + k_{z2})\eta\sin\alpha(1-\cos i_T)$ $\qquad a_{1,2}^x = 0 \quad b_{1,2}^x = 1 \quad \theta_{1,2}^x = \varphi_1$

$C_{1,3}^x = -\frac{1}{4}(\dot{z}_0 + k_{z2})\eta\sin\alpha(1+\cos i_T)$ $\qquad a_{1,3}^x = 0 \quad b_{1,3}^x = 1 \quad \theta_{1,3}^x = -\varphi_2$

$$\begin{aligned}C_{1,4}^x = &-\frac{1}{2}\frac{c_1(n_T+\omega_E)}{\omega_E(\omega_E+2n_T)}\eta\cos\alpha\sin i_T + \frac{1}{2}\frac{c_2(n_T-\omega_E)}{\omega_E(\omega_E-2n_T)}\eta\cos\alpha\sin i_T\\ &+\frac{1}{2}\frac{c_3}{4\omega_T}\eta\sin\alpha\cos i_T - R_T\eta\sin\alpha\left[\sin i_T(n_T - \omega_E\cos i_T) + \frac{1}{4}\sin 2i_T\omega_E\right]\end{aligned}$$

$a_{1,4}^x = 0 \quad b_{1,4}^x = 1 \quad \theta_{1,4}^x = \psi_1$

© Springer Science+Business Media Singapore 2017

Y. Yan et al., *Dynamics and Control of Lorentz-Augmented Spacecraft Relative Motion*, DOI 10.1007/978-981-10-2603-4

$$C^x_{1,5} = \tfrac{1}{4}(\dot{z}_0 + k_{z2})\eta \sin\alpha(1 - \cos i_T) \qquad a^x_{1,5} = 2 \quad b^x_{1,5} = 1 \quad \theta^x_{1,5} = \varphi_1$$

$$C^x_{1,6} = \tfrac{1}{4}(\dot{z}_0 + k_{z2})\eta \sin\alpha(1 + \cos i_T) \qquad a^x_{1,6} = 2 \quad b^x_{1,6} = -1 \quad \theta^x_{1,6} = \varphi_2$$

$$C^x_{1,7} = -\frac{1}{2}\frac{c_1(n_T + \omega_E)}{\omega_E(\omega_E + 2n_T)}\eta \cos\alpha \sin i_T - \frac{1}{4}\frac{c_3}{4n_T}\eta \sin\alpha(1 - \cos i_T) + \frac{1}{8}R_T\eta \sin\alpha(2\sin i_T - \sin 2i_T)\omega_E \qquad a^x_{1,7} = 2 \quad b^x_{1,7} = 1 \quad \theta^x_{1,7} = \psi_2$$

$$C^x_{1,8} = -\frac{1}{2}\frac{c_2(n_T - \omega_E)}{\omega_E(\omega_E - 2n_T)}\eta \cos\alpha \sin i_T - \frac{1}{4}\frac{c_3}{4n_T}\eta \sin\alpha(1 + \cos i_T) + \frac{1}{8}R_T\eta \sin\alpha(2\sin i_T + \sin 2i_T)\omega_E \qquad a^x_{1,8} = 2 \quad b^x_{1,8} = -1 \quad \theta^x_{1,8} = \psi_3$$

$$C^x_{2,1} = -\tfrac{1}{2}\tfrac{c_3}{2}\eta \cos\alpha \sin i_T \qquad a^x_{2,1} = 2 \quad b^x_{2,1} = 0 \quad \theta^x_{2,1} = 2\varphi_3$$

$$C^x_{1,9} = 0, C^x_{2,i} = 0 (i \geq 2)$$

$$D^x_{1,1} = -\tfrac{1}{2}(\dot{z}_0 + k_{z2})\eta \cos\alpha \sin i_T \qquad c^x_{1,1} = 2 \quad d^x_{1,1} = 0 \quad \phi^x_{1,1} = \varphi_3$$

$$D^x_{1,2} = -\frac{1}{2}R_T\eta \cos\alpha \sin^2 i_T\omega_E - \frac{1}{4}\frac{c_1(n_T + \omega_E)}{\omega_E(\omega_E + 2n_T)}\eta \sin\alpha(1 + \cos i_T) - \frac{1}{4}\frac{c_2(n_T - \omega_E)}{\omega_E(\omega_E - 2n_T)}\eta \sin\alpha(1 - \cos i_T) + \frac{1}{2}\frac{c_3}{4n_T}\eta \cos\alpha \sin i_T \qquad c^x_{1,2} = 2 \quad d^x_{1,2} = 0 \quad \phi^x_{1,2} = 2\varphi_3$$

$$D^x_{1,3} = -\tfrac{1}{4}(z_0 + k_{z1})\eta \sin\alpha(1 - \cos i_T)n_T \qquad c^x_{1,3} = 0 \quad d^x_{1,3} = 1 \quad \phi^x_{1,3} = \varphi_1$$

$$D^x_{1,4} = -\tfrac{1}{4}(z_0 + k_{z1})\eta \sin\alpha(1 + \cos i_T)n_T \qquad c^x_{1,4} = 0 \quad d^x_{1,4} = 1 \quad \phi^x_{1,4} = -\varphi_2$$

$$D^x_{1,5} = \frac{1}{4}\frac{c_1(n_T + \omega_E)}{\omega_E(\omega_E + 2n_T)}\eta \sin\alpha(1 + \cos i_T) + \frac{1}{4}\frac{c_2(n_T - \omega_E)}{\omega_E(\omega_E - 2n_T)}\eta \sin\alpha(1 - \cos i_T) \qquad c^x_{1,5} = 0 \quad d^x_{1,5} = 2 \quad \phi^x_{1,5} = 2\psi_1$$

$$D^x_{1,6} = \tfrac{1}{4}(z_0 + k_{z1})\eta \sin\alpha(1 - \cos i_T)n_T \qquad c^x_{1,6} = 2 \quad d^x_{1,6} = 1 \quad \phi^x_{1,6} = \varphi_1$$

$$D^x_{1,7} = -\tfrac{1}{4}\tfrac{c_1(n_T + \omega_E)}{\omega_E(\omega_E + 2n_T)}\eta \sin\alpha(1 - \cos i_T) \qquad c^x_{1,7} = 2 \quad d^x_{1,7} = 2 \quad \phi^x_{1,7} = 2\varphi_1$$

$$D^x_{1,8} = \tfrac{1}{4}(z_0 + k_{z1})\eta \sin\alpha(1 + \cos i_T)n_T \qquad c^x_{1,8} = 2 \quad d^x_{1,8} = -1 \quad \phi^x_{1,8} = \varphi_2$$

$$D^x_{1,9} = -\tfrac{1}{4}\tfrac{c_2(n_T - \omega_E)}{\omega_E(\omega_E - 2n_T)}\eta \sin\alpha(1 + \cos i_T) \qquad c^x_{1,9} = 2 \quad d^x_{1,9} = -2 \quad \phi^x_{1,9} = 2\varphi_2$$

$$D^x_{2,1} = \tfrac{1}{2}\tfrac{c_3}{2}\eta \sin\alpha \qquad c^x_{2,1} = 0 \quad d^x_{2,1} = 1 \quad \phi^x_{2,1} = \psi_1$$

$D^x_{2,2} = -\frac{1}{4}\frac{c_3}{2}\eta \sin\alpha(1-\cos i_T)$ $\quad c^x_{2,2}=2 \quad d^x_{2,2}=1 \quad \phi^x_{2,2}=\psi_2$

$D^x_{2,3} = -\frac{1}{4}\frac{c_3}{2}\eta \sin\alpha(1+\cos i_T)$ $\quad c^x_{2,3}=2 \quad d^x_{2,3}=-1 \quad \phi^x_{2,3}=\psi_3$

$D^x_{2,i} = 0(i \geq 4)$

$C^y_{1,1} = -(\dot{z}_0+k_{z2})\eta \cos\alpha \sin i_T$ $\quad a^y_{1,1}=2 \quad b^y_{1,1}=0 \quad \theta^y_{1,1}=\varphi_3$

$$C^y_{1,2} = -R_T\eta \cos\alpha \sin^2 i_T\omega_E - \frac{1}{2}\frac{c_1(n_T+\omega_E)}{\omega_E(\omega_E+2n_T)}\eta \sin\alpha(1+\cos i_T) - \frac{1}{2}\frac{c_2(n_T-\omega_E)}{\omega_E(\omega_E-2n_T)}\eta \sin\alpha(1-\cos i_T) + \frac{1}{2}\frac{c_3}{2n_T}\eta \cos\alpha \sin i_T$$

$a^y_{1,2}=2 \quad b^y_{1,2}=0 \quad \theta^y_{1,2}=2\varphi_3$

$C^y_{1,3} = -\frac{1}{2}(z_0+k_{z1})\eta \sin\alpha(1-\cos i_T)n_T$ $\quad a^y_{1,3}=0 \quad b^y_{1,3}=1 \quad \theta^y_{1,3}=\varphi_1$

$C^y_{1,4} = \frac{1}{2}(z_0+k_{z1})\eta \sin\alpha(1+\cos i_T)n_T$ $\quad a^y_{1,4}=0 \quad b^y_{1,4}=1 \quad \theta^y_{1,4}=-\varphi_2$

$C^y_{1,5} = -\frac{1}{2}\frac{c_1(n_T+\omega_E)}{\omega_E(\omega_E+2n_T)}\eta \sin\alpha(1+\cos i_T) + \frac{1}{2}\frac{c_2(n_T-\omega_E)}{\omega_E(\omega_E-2n_T)}\eta \sin\alpha(1-\cos i_T)$ $\quad a^y_{1,5}=0 \quad b^y_{1,5}=2 \quad \theta^y_{1,5}=2\psi_1$

$C^y_{1,6} = \frac{1}{2}(z_0+k_{z1})\eta \sin\alpha(1-\cos i_T)n_T$ $\quad a^y_{1,6}=2 \quad b^y_{1,6}=1 \quad \theta^y_{1,6}=\varphi_1$

$C^y_{1,7} = -\frac{1}{2}\frac{c_1(n_T+\omega_E)}{\omega_E(\omega_E+2n_T)}\eta \sin\alpha(1-\cos i_T)$ $\quad a^y_{1,7}=2 \quad b^y_{1,7}=2 \quad \theta^y_{1,7}=2\varphi_1$

$C^y_{1,8} = \frac{1}{2}(z_0+k_{z1})\eta \sin\alpha(1+\cos i_T)n_T$ $\quad a^y_{1,8}=2 \quad b^y_{1,8}=-1 \quad \theta^y_{1,8}=\varphi_2$

$C^y_{1,9} = -\frac{1}{2}\frac{c_2(n_T-\omega_E)}{\omega_E(\omega_E-2n_T)}\eta \sin\alpha(1+\cos i_T)$ $\quad a^y_{1,9}=2 \quad b^y_{1,9}=-2 \quad \theta^y_{1,9}=2\varphi_2$

$C^y_{2,1} = -\frac{c_3}{2}\eta \sin\alpha \cos i_T$ $\quad a^y_{2,1}=0 \quad b^y_{2,1}=1 \quad \theta^y_{2,1}=\psi_1$

$C^y_{2,2} = -\frac{1}{2}\frac{c_3}{2}\eta \sin\alpha(1-\cos i_T)$ $\quad a^y_{2,2}=2 \quad b^y_{2,2}=1 \quad \theta^y_{2,2}=\psi_2$

$C^y_{2,3} = -\frac{1}{2}\frac{c_3}{2}\eta \sin\alpha(1+\cos i_T)$ $\quad a^y_{2,3}=2 \quad b^y_{2,3}=-1 \quad \theta^y_{2,3}=\psi_3$

$C^y_{2,i} = 0(i \geq 4)$

$D^y_{1,1} = -(z_0+k_{z1})\eta \cos\alpha \sin i_T n_T$ $\quad c^y_{1,1}=2 \quad d^y_{1,1}=0 \quad \phi^y_{1,1}=\varphi_3$

$D^y_{1,2} = -\frac{1}{2}(\dot{z}_0+k_{z2})\eta \sin\alpha(1-\cos i_T)$ $\quad c^y_{1,2}=0 \quad d^y_{1,2}=1 \quad \phi^y_{1,2}=\varphi_1$

$D^y_{1,3} = -\frac{1}{2}(\dot{z}_0+k_{z2})\eta \sin\alpha(1+\cos i_T)$ $\quad c^y_{1,3}=0 \quad d^y_{1,3}=1 \quad \phi^y_{1,3}=-\varphi_2$

$$D^y_{1,4} = -\frac{c_1(n_T+\omega_E)}{\omega_E(\omega_E+2n_T)}\eta \cos\alpha \sin i_T - \frac{c_2(n_T-\omega_E)}{\omega_E(\omega_E-2n_T)}\eta \cos\alpha \sin i_T + \frac{c_3}{4\omega_T}\eta \sin\alpha - R_T\eta \sin\alpha \sin i_T\omega_E$$

$c^y_{1,4}=0 \quad d^y_{1,4}=1 \quad \phi^y_{1,4}=\psi_1$

$D^y_{1,5} = -\frac{1}{2}(\dot{z}_0+k_{z2})\eta \sin\alpha(1-\cos i_T)$ $\quad c^y_{1,5}=2 \quad d^y_{1,5}=1 \quad \phi^y_{1,5}=\varphi_1$

$D^y_{1,6} = -\frac{1}{2}(\dot{z}_0+k_{z2})\eta \sin\alpha(1+\cos i_T)$ $\quad c^y_{1,6}=2 \quad d^y_{1,6}=-1 \quad \phi^y_{1,6}=\varphi_2$

$$D^{y}_{1,7} = \frac{c_1(n_T+\omega_E)}{\omega_E(\omega_E+2n_T)}\eta\cos\alpha\sin i_T + \frac{1}{2}\frac{c_3}{4n_T}\eta\sin\alpha(1-\cos i_T) \qquad c^{y}_{1,7}=2 \quad d^{y}_{1,7}=1 \quad \phi^{y}_{1,7}=\psi_2$$
$$-\frac{1}{2}R_T\eta\sin\alpha(1-\cos i_T)\sin i_T\omega_E$$

$$D^{y}_{1,8} = \frac{c_2(n_T-\omega_E)}{\omega_E(\omega_E-2n_T)}\eta\cos\alpha\sin i_T + \frac{1}{2}\frac{c_3}{4n_T}\eta\sin\alpha(1+\cos i_T) \qquad c^{y}_{1,8}=2 \quad d^{y}_{1,8}=-1 \quad \phi^{y}_{1,8}=\psi_3$$
$$-\frac{1}{2}R_T\eta\sin\alpha(1+\cos i_T)\sin i_T\omega_E$$

$$D^{y}_{2,1} = \tfrac{1}{2}c_3\eta\cos\alpha\sin i_T \qquad c^{y}_{2,1}=2 \quad d^{y}_{2,1}=0 \quad \phi^{y}_{1,1}=2\varphi_3$$

$$D^{y}_{1,9}=0, D^{y}_{2,i}=0(i\geq 2) \tag{A.6}$$

where

$$\begin{aligned}\psi_1 &= \psi_M-\Omega_T\\ \psi_2 &= 2u_{T0}+\psi_M-\Omega_T\\ \psi_3 &= 2u_{T0}-\psi_M+\Omega_T\\ \psi_4 &= 2u_{T0}\end{aligned} \tag{A.7}$$

A.2 Expressions of λ_x and λ_y

The expression of $\lambda_x(t)$ is given by

$$\begin{aligned}\lambda_x(t) =& k_{x1}\cos n_T t + \frac{1}{n_T}k_{x2}\sin n_T t + \frac{1}{n_T^2}\left(\tilde{\kappa}^{x}_{t,0}+\tilde{\kappa}^{x}_{t,1}t+\tilde{\kappa}^{x}_{t,2}t^2\right) - \frac{2}{n_T^4}\tilde{\kappa}^{x}_{t,2}\\ &+\sum_{m=1}^{2}t^{m-1}\left[\sum_{i=1}^{9}\frac{\widetilde{C}^{x}_{m,i}}{n_T^2-\left(A^{x}_{m,i}\right)^2}\sin\left(A^{x}_{m,i}t+\theta^{x}_{m,i}\right)+\frac{\widetilde{D}^{x}_{m,i}}{n_T^2-\left(B^{x}_{m,i}\right)^2}\cos\left(B^{x}_{m,i}t+\phi^{x}_{m,i}\right)\right.\\ &+\sum_{i=1}^{9}\frac{\widetilde{P}^{x}_{2,i}}{n_T^2-\left(B^{x}_{2,i}\right)^2}\sin\left(B^{x}_{2,i}t+\phi^{x}_{2,i}\right)+\frac{\widetilde{Q}^{x}_{2,i}}{n_T^2-\left(A^{x}_{2,i}\right)^2}\cos\left(A^{x}_{2,i}t+\theta^{x}_{2,i}\right)\end{aligned} \tag{A.8}$$

with

$$k_{x1} = -\frac{1}{n_T^4}\left(n_T^2\tilde{\kappa}_{t,0}^x - 2\tilde{\kappa}_{t,2}^x\right) - \sum_{i=1}^{9}\left[\frac{\widetilde{C}_{1,i}^x}{n_T^2 - \left(A_{1,i}^x\right)^2}\sin\theta_{1,i}^x + \frac{\widetilde{D}_{1,i}^x}{n_T^2 - \left(B_{1,i}^x\right)^2}\cos\phi_{1,i}^x\right.$$
$$\left. + \frac{\widetilde{P}_{2,i}^x}{n_T^2 - \left(B_{2,i}^x\right)^2}\sin\phi_{2,i}^x + \frac{\widetilde{Q}_{2,i}^x}{n_T^2 - \left(A_{2,i}^x\right)^2}\cos\theta_{2,i}^x\right] \tag{A.9}$$

$$k_{x2} = -\frac{\tilde{\kappa}_{t,1}^x}{n_T^2} - \sum_{i=1}^{9}\left[\frac{\widetilde{C}_{1,i}^x A_{1,i}^x}{n_T^2 - \left(A_{1,i}^x\right)^2}\cos\theta_{1,i}^x - \frac{\widetilde{D}_{1,i}^x B_{1,i}^x}{n_T^2 - \left(B_{1,i}^x\right)^2}\sin\phi_{1,i}^x\right.$$
$$\left. + \frac{\widetilde{C}_{2,i}^x - \widetilde{Q}_{2,i}^x A_{2,i}^x}{n_T^2 - \left(A_{2,i}^x\right)^2}\sin\theta_{2,i}^x + \frac{\widetilde{D}_{2,i}^x + \widetilde{P}_{2,i}^x B_{2,i}^x}{n_T^2 - \left(B_{2,i}^x\right)^2}\cos\phi_{2,i}^x\right] \tag{A.10}$$

$$\tilde{\kappa}_{t,0}^x = \kappa^y + \kappa_{t,0}^x, \quad \tilde{\kappa}_{t,1}^x = 2n_T\kappa_{t,0}^y, \quad \tilde{\kappa}_{t,2}^x = n_T\kappa_{t,1}^y \tag{A.11}$$

$$\widetilde{C}_{m,i}^x = C_{m,i}^x + 2n_T\frac{D_{m,i}^y}{B_{m,i}^y}, \quad \widetilde{D}_{m,i}^x = D_{m,i}^x - 2n_T\frac{C_{m,i}^y}{A_{m,i}^y} \tag{A.12}$$

$$\widetilde{P}_{2,i}^x = P_{2,i}^x + \frac{2\tilde{D}_{2,i}^x B_{2,i}^x}{n_T^2 - \left(B_{2,i}^x\right)^2}, \quad \widetilde{Q}_{2,i}^x = Q_{2,i}^x - \frac{2\tilde{C}_{2,i}^x A_{2,i}^x}{n_T^2 - \left(A_{2,i}^x\right)^2} \tag{A.13}$$

where

$$\kappa^y = -2n_T\sum_{i=1}^{9}\left[-\frac{C_{1,i}^y}{A_{1,i}^y}\cos\theta_{1,i}^y + \frac{C_{2,i}^y}{\left(A_{2,i}^y\right)^2}\sin\theta_{2,i}^y\right.$$
$$\left. + \frac{D_{1,i}^y}{B_{1,i}^y}\sin\phi_{1,i}^y + \frac{D_{2,i}^y}{\left(B_{2,i}^y\right)^2}\cos\phi_{2,i}^y\right] \tag{A.14}$$

$$P_{2,i}^x = 2n_T\frac{C_{2,i}^y}{\left(A_{2,i}^y\right)^2}, \quad Q_{2,i}^x = 2n_T\frac{D_{2,i}^y}{\left(B_{2,i}^y\right)^2} \tag{A.15}$$

The expression of $\lambda_y(t)$ is given by

$$\lambda_y(t) = \frac{1}{n_T} k_{y1} \cos n_T t + k_{y2} \sin n_T t + \tilde{\kappa}^y_{t,0} + \tilde{\kappa}^y_{t,1} t + \tilde{\kappa}^y_{t,2} t^2 + \tilde{\kappa}^y_{t,3} t^3$$
$$+ \sum_{m=1}^{2} t^{m-1} \left[\sum_{i=1}^{9} \frac{\widetilde{C}^y_{m,i}}{n_T^2 - \left(A^y_{m,i}\right)^2} \sin\left(A^y_{m,i} t + \theta^y_{m,i}\right) + \frac{\widetilde{D}^y_{m,i}}{n_T^2 - \left(B^y_{m,i}\right)^2} \cos\left(B^y_{m,i} t + \phi^y_{m,i}\right) \right]$$
$$+ \sum_{i=1}^{9} \frac{\widetilde{P}^y_{2,i}}{n_T^2 - \left(B^y_{2,i}\right)^2} \sin\left(B^y_{2,i} t + \phi^y_{2,i}\right) + \frac{\widetilde{Q}^y_{2,i}}{n_T^2 - \left(A^y_{2,i}\right)^2} \cos\left(A^y_{2,i} t + \theta^y_{2,i}\right) \tag{A.16}$$

where

$$k_{y1} = 2k_{x2}, \quad k_{y2} = -2k_{x1} \tag{A.17}$$

$$\tilde{\kappa}^y_{t,0} = -\frac{1}{n_T} k_{y1} - \sum_{i=1}^{9} \left[\frac{\widetilde{C}^y_{1,i}}{n_T^2 - \left(A^y_{1,i}\right)^2} \sin \theta^y_{1,i} + \frac{\widetilde{D}^y_{1,i}}{n_T^2 - \left(B^y_{1,i}\right)^2} \cos \phi^y_{1,i} \right.$$
$$\left. + \frac{\widetilde{P}^y_{2,i}}{n_T^2 - \left(B^y_{2,i}\right)^2} \sin \phi^y_{2,i} + \frac{\widetilde{Q}^y_{2,i}}{n_T^2 - \left(A^y_{2,i}\right)^2} \cos \theta^y_{2,i} \right] \tag{A.18}$$

$$\tilde{\kappa}^y_{t,1} = \frac{\kappa^y}{2n_T} + \frac{4}{n_T^3} \tilde{\kappa}^x_{t,2} - \frac{2}{n_T} \tilde{\kappa}^x_{t,0}, \quad \tilde{\kappa}^y_{t,2} = -\frac{3}{2} \kappa^y_{t,0}, \quad \tilde{\kappa}^y_{t,3} = -\frac{1}{2} \kappa^y_{t,1} \tag{A.19}$$

$$\widetilde{C}^y_{m,i} = \left\{ -2n_T \tilde{D}^x_{m,i} - \frac{C^y_{m,i}}{A^y_{m,i}} \left[n_T^2 - \left(A^y_{m,i}\right)^2 \right] \right\} \frac{1}{A^y_{m,i}} \tag{A.20}$$

$$\widetilde{D}^y_{m,i} = \left\{ 2n_T \tilde{C}^x_{m,i} - \frac{D^y_{m,i}}{B^y_{m,i}} \left[n_T^2 - \left(B^y_{m,i}\right)^2 \right] \right\} \frac{1}{B^y_{m,i}} \tag{A.21}$$

$$\widetilde{P}^y_{2,i} = \left\{ -2n_T \left(\tilde{C}^x_{2,i} + \tilde{Q}^x_{2,i} B^y_{2,i} \right) + 2 \frac{D^y_{2,i}}{B^y_{2,i}} \left[n_T^2 - \left(B^y_{2,i}\right)^2 \right] \right\} \frac{1}{\left(B^y_{2,i}\right)^2} \tag{A.22}$$

$$\widetilde{Q}^y_{2,i} = \left\{ -2n_T \left(\tilde{D}^x_{2,i} - \tilde{P}^x_{2,i} A^y_{2,i} \right) - 2 \frac{C^y_{2,i}}{A^y_{2,i}} \left[n_T^2 - \left(A^y_{2,i}\right)^2 \right] \right\} \frac{1}{\left(A^y_{2,i}\right)^2} \tag{A.23}$$

Printed in the United States
By Bookmasters